Masks of the Universe

Changing Ideas on the Nature of the Cosmos

To the ancient Greeks the universe consisted of earth, air, fire, and water. To Saint Augustine it was the Word of God. To many modern scientists it is the dance of atoms and waves, and in years to come it may be different again. What then is the real Universe? History shows that in every age each society constructs its own universe, believing it to be the real and final Universe. Yet each universe is only a model or mask of the unknown Universe. This book brings together fundamental scientific, philosophical, and religious issues in cosmology, raising thought-provoking questions. In every age people have pitied the universes of their ancestors, convinced that they have at last discovered the ultimate truth. Do we now stand at the threshold of knowing everything, or will our latest model, like all the rest, be pitied by our descendants?

Edward Harrison is Emeritus Distinguished Professor of Physics and Astronomy at the University of Massachusetts, and adjunct Professor of Astronomy at the Steward Observatory, University of Arizona. He was born and educated in England, and served for several years in the British Army during World War II. He was principal scientist at the Atomic Energy Research Establishment and Rutherford High Energy Laboratory until 1966, when he became a Five College Professor at the University of Massachusetts, and taught at Amhert, Hampshire, Mount Holyoke, and Smith Colleges. He has written several books, including *Cosmology: The Science of the Universe*, also published by Cambridge University Press, and has published hundreds of technical papers in physics and astronomy journals.

Masks of the Universe

Changing Ideas on the Nature of the Cosmos

EDWARD HARRISON

University of Arizona

Second edition

CAMBRIDGE
UNIVERSITY PRESS

PUBLISHED BY THE PRESS SYNDICATE OF THE UNIVERSITY OF CAMBRIDGE
The Pitt Building, Trumpington Street, Cambridge, United Kingdom

CAMBRIDGE UNIVERSITY PRESS
The Edinburgh Building, Cambridge CB2 2RU, UK
40 West 20th Street, New York, NY 10011-4211, USA
477 Williamstown Road, Port Melbourne, VIC 3207, Australia
Ruiz de Alarcón 13, 28014 Madrid, Spain
Dock House, The Waterfront, Cape Town 8001, South Africa

http://www.cambridge.org

First edition published 1985
Second edition published 2003

Printed in the United Kingdom at the University Press, Cambridge

Typeface Trump Mediaeval 9.5/15 pt   *System* LaTeX 2$_\varepsilon$ [TB]

A catalog record for this book is available from the British Library

Library of Congress Cataloging in Publication data

Harrison, Edward Robert.
Masks of the universe / Edward Harrison.–2nd ed.
    p. cm.
Includes bibliographical references and index.
ISBN 0 521 77351 2
1. Cosmology.   I. Title.
QB981.H324   2003
523.1–dc21   2002031342

ISBN 0 521 77351 2 hardback

Contents

Preface

In the preface to the first edition of *Masks of Universe* I wrote:
"At first I thought this book would take me only a few months to
write. After all, the basic idea was simple, and only a few words
should suffice to make it clear and convincing. But soon this illusion
was shattered. A few months grew into three years, and now I realize
that thirty years would not suffice. But enough! Other work presses,
and life is too short." Here I am, not thirty years but almost two
decades later writing the preface to the second edition and struggling
again to make clear the "simple idea."

The idea rests on the distinction between Universe and
universes. The Universe by definition is everything and includes us
experiencing and thinking about it. The universes are the models of
the Universe that we construct to explain our observations and
experiences. Beneath the deceptive simplicity of this idea lies
a little-explored realm of thought.

No person can live in a society of intelligent members unless
equipped with grand ideas of the world around. These grand
ideas – or cosmic formulations – establish the universe in which that
society lives. The universes that human beings devise and in which
they live, or believe they live, organize and give meaning to their
experiences. Where there is a society of intelligent beings (not
necessarily intelligent by our standards), there we find a rational
universe (not necessarily rational by our standards); where there is
a universe, there we find a society. The universes are the masks of
the Universe. The unmasked Universe itself, however, remains
forever beyond full human comprehension.

The Universe is everything and includes us thinking about it.
We are, in fact, the Universe thinking about itself. How can we, who

are only a very limited part or aspect, comprehend the whole? Modesty alone suggests we cannot in any absolute sense. We comprehend instead a universe that we have ourselves conceptually devised: a model of the unknown Universe.

History shows that the Universe is patient of many interpretations. Each interpretation is a model – a universe – a mask fitted on the faceless Universe. Every human society has its universe. The Egyptian, Babylonian, Zoroastrian, Aristotelian, Epicurean, Stoic, Neoplatonic, Medieval, Newtonian, Victorian universes are examples.

Each universe in its day stands as an awe-inspiring "reality," yet each is doomed to be superseded by another and perhaps grander "reality." Each is a framework of concepts that explains what is observed and determines what is significant. Each organizes human experience and shapes human thought. The members of a society believe in the truth of their universe and mistake it always for the Universe. Prophets proclaim it, religions authenticate it, empires glorify it, and wars promote it. In each universe the end of knowledge looms in sight. Always only a few things remain to be discovered. We pity the universes of our ancestors and forget that our descendants will pity us for the same reason.

In cosmology in the ancient world philosophical issues dominated. In the Middle Ages theological issues ranked foremost. In recent times astronomy and the physical sciences have taken over and philosophical issues concerning the cosmos now receive scant attention. Yet the clear articulations of modern science have brought into sharper focus than ever before still unresolved philosophical and theological problems.

For example, consider the containment riddle (see *Cosmology: The Science of the Universe*). The current universe (actually any universe), which supposedly is all-inclusive, contains us contemplating that particular universe. But this leads into an infinite regression: the universe contains us contemplating the universe that contains us contemplating the universe that contains..., and so on, indefinitely. The riddle is solved by stressing the distinction between Universe and universe. Thus: The Universe, which by definition is all-inclusive, contains us contemplating the current universe. There is now no regression for the image does not contain the image-maker. The universe contains only representations of us in the form of bodies and brains, whereas our contemplative minds with their consciousness and free will are of the Universe and make no substantial and explicit contribution to the makeup of our deterministic universes. What is not contained in a universes is not necessarily nonexistent.

The new edition is mostly rewritten and includes two new chapters, one on time (tentatively foreseeing possible future changes in our understanding of time), and the other on the *ultimum sentiens* (a study of who or what actually does the perceiving).

I am grateful to the Institute of Astronomy, Cambridge University, for hospitality, and the University of Massachusetts for a Faculty Fellowship that enabled me to complete the first edition. I am grateful to literally hundreds of people for their valuable comments, and also I am indebted to many old friends, including Vere Chappel, John Roberts, Carl Swanson, Oswald Tippo, and Peter Webster for their comments on certain ideas, and to Michael Arbib, Thomas Arny, Leroy Cook, Jay Demerath, Seymour Epstein, Laurence Marschall, Gordon Sutton, David Van Blerkom, and Richard Ziemacki for their helpful comments on various chapters. Finally, I acknowledge gratefully the insightful comments made by my wife Photeni, son Peter, and daughter June Harrison.

I   Introducing the Masks

The theme of this book is that the universe in which we live, or think we live, is mostly a thing of our own making. The underlying idea is the distinction between *Universe* and *universes*. It is a simple idea having many consequences.

The Universe is everything. What it is, in its own right, independent of our changing opinions, we never fully know. It is all-inclusive and includes us as conscious beings. We are a part or an aspect of the Universe experiencing and thinking about itself.

What is the Universe? Seeking an answer is the endless quest. I can think of no better reply than the admission by Socrates: "all that I know is that I know nothing." David Hume, a Scottish philosopher in the eighteenth century, in reply to a similar question, said "it admits of no answer" for absolute truth is inaccessible to the human mind. Logan Smith, an expatriate American living in London, expressed his reply in a witty essay *Trivia* (1902), "I awoke this morning . . . into the daylight, the furniture of my bedroom – in fact, into the well-known, often-discussed, but to my mind as yet unexplained Universe."

The universes are our models of the Universe. They are great schemes of intricate thought – grand belief systems – that rationalize the human experience. They harmonize and invest with meaning the rising and setting Sun, the waxing and waning Moon, the jeweled lights of the night sky, the landscapes of rocks and trees, and the tumult of everyday life. Each determines what is perceived and what constitutes valid knowledge, and the members of a society believe what they perceive and perceive what they believe. *A universe is a mask fitted on the face of the unknown Universe.*

*   *   *

Where there is a society of human beings, however primitive, there we find a universe; and where there is a universe, of whatever kind, there we find a society. Both go together, the one does not exist without the other. A universe unifies a society, enabling its members to communicate and share their thoughts and experiences. A universe might not be rational by our standards, or those of other societies, but is always rational by the standards of its own society. Our universe, the universe in which we live, or think we live, is the modern physical universe.

The conscious mind with its sense of free will belongs to the Universe; the physical brain with its neurological structures belongs to the physical universe. By failing to recognize the difference between Universe and universe, and by believing that the physical universe is the Universe, we are left stranded with no recourse other than to discard mind and freewill as fictional hangovers from past belief systems. They have no place in the physical scheme of things, and in the natural sciences we consciously deny the existence of consciousness.

The Universe is everything and includes us struggling to understand it by devising representative universes. One might say the universes are the Universe seeking to understand itself. René Descartes, a philosopher in the seventeenth century, doubting everything except the existence of his doubts, announced "I doubt, therefore I think. I think, therefore I am." The reality of everything else was left in doubt. He saved the day by invoking God as an infallible arbiter of reliable truth. An alternative and more inclusive ontological argument might state, "I think, therefore I am. I am part of the Universe, therefore the Universe thinks. The Universe thinks, therefore it is." To doubt the Universe, is to doubt our own existence.

Friedrich Nietzsche, a German philosopher of the mid-nineteenth century, said "God is dead," and like many others despaired of the universe having any ultimate meaning. But like others he confused the universe that he thought he lived in with the Universe. Albert Einstein, foremost twentieth century scientist, once said: "The most incomprehensible thing about the universe is that it is comprehensible." We may complement Einstein's remark by adding: "The most comprehensible thing about the Universe is that it is incomprehensible." A universe – any universe – is comprehensible because it has been shaped by the human mind. Whereas the Universe is incomprehensible if only because we can never grasp the entirety of a reality of which we are only a part or an aspect. The Universe may comprehend itself, but not by means of finite human minds.

*   *   *

Cosmology is the study of universes. It is a prodigious enterprise embracing all branches of knowledge. Naturally, cosmologists occupy themselves primarily with the study of the contemporary universe. One universe at a time is more than enough. Why bother with the universes of the past when they were all wrong? Why try to anticipate the universes of the future when the present universe, apart from a few loose ends, is already the correct and final model?

The realization that universes are impermanent conceptual schemes comes from the study of history. This aspect of cosmology is rarely stressed and might come as a surprise. Automatically, we tend to regard the universes of earlier societies as pathetically unreal in comparison with our own. It is disconcerting to be told that our modern physical universe is the latest model that almost certainly in the future will be discarded and replaced with another and possibly more resplendent model.

We cannot understand our universe and see it in full perspective without heeding the earlier universes from which it springs. Through the historian's eyes we see the past as a gallery of grand cosmic pictures, and we wonder, is our universe the final picture, have we arrived at last at the end of the gallery? We see the past as a procession of masks – masks of awesome grandeur – and we wonder, will the procession continue endlessly into the future? And if there is no end in sight to the gallery of pictures, no end to the mockery of masks, what are we to make of the contemporary universe in which we live, or think we live? This book is my search for an answer.

*   *   *

Throughout history the end of knowledge has always loomed in sight. A few things always remain to be discovered, a few problems to be solved, then everything will be crystal clear. Either we shall have attained the throne of God, acquired the philosopher's stone, genetically reinvented ourselves, explored other star systems, discovered extraterrestrial life, converted everybody to our own brand of religion, made global our political system, or found the theory that explains everything. Always this or that subject of burning interest is said to be the final frontier. Pity the people of the future! What will they do when all knowledge has been discovered? This oldest of human conceits, which confuses universe with Universe, is alive today as much as at any time in the past. We are afflicted with the hubris that denies our descendants the right to different and better knowledge.

As a society evolves, its universe also develops and evolves. Then, within an ace of understanding everything, the old universe dissolves in a ferment of social upheaval and a new universe emerges, full of promise and exciting challenge. Universes rise, flourish for a decade, a century, or a millennium, and decline. They decline because of the assault of an alien culture, or revolutionary ideas refuse to remain suppressed, or old problems reappear and take center stage, or for no other reason than the climate of opinion changes.

*   *   *

Often we pretend not to live in the universe, knowing that we pretend. We alternate between no pretense, when we live in the "real" world of our society, and double pretense when we pretend to live in a pretended world and "all that we see or seem is but a dream within a dream." It is the natural way a sane person lives. We withdraw into counterfeit worlds of fiction and fantasy when the reality of the universe becomes too much. On returning, we put down the book, turn off the television, come home from the play, feeling entertained, knowing that we have lived in a counterfeit world.

But those individuals lost and tragically betrayed by the universe, who cannot alternate between no pretense and double pretense, who find sanctuary in a private world of pretense, unaware of its pretense, they, we deem, are the insane.

But what of the universes that betray not just a few but most members of their societies? These are the mad universes created and ruled by sick minds. In the annals of history they are many. We may mention, as examples, the witch universe that terrorized the Renaissance, the pathological universes of societies engaged in bitter religious and political wars, and the oppressive universes of totalitarian societies. Mad universes impose termite uniformity, suppress freedom, exalt the authority of the state, rule by fear, and often, but not always, are blessedly short-lived. Sooner or later the societies of mad universes are eliminated by the intricate processes of natural selection.

*   *   *

In the garden, as I write, hosts of golden daffodils are fluttering and dancing in the breeze. You and I live in that world out there of hills, lakes, trees, and daffodils with its multitude of things and torrent of events, and the overarching picture we share is the physical universe.

Most of us understand very little about the physical universe, about atoms, cells, and stars. Some of us may even dislike the physical universe. But unlike the members of earlier societies, we drive automobiles while listening to the radio, communicate worldwide by internet and telephone, fly in planes to distant lands, watch television, use computers, depend on modern medicine, and use electricity in a myriad ways. We may not understand the physical universe, and we may not like it, but we depend on it, and we believe in it. Only an insane person totally disbelieves in the physical universe.

People in earlier societies had other outlooks. The Babylonians, Egyptians, Minoans, Ionians, Mayans, Iroquois, Maori,..., lived in universes all different and none was like the modern physical universe. In the Babylonian universe the flowers danced and fluttered in the breeze, the Sun rose and set, the Moon waxed and waned, the constellations wheeled across the night sky, and a rock was a rock and a tree a tree. But the meaning of these things was greatly different from what we now deem is natural. The Babylonian, Egyptian, ... universes, so unlike our own, were in harmony with the cultures and modes of thought of their societies.

Common sense tells us that the out-of-date and discarded universes of the past, going back hundreds of thousands of years, were all much mistaken in their general and detailed view of things. But, and here comes the rub, it does not take much thought to realize that the people in the past believed in their universes, just as strongly as we now believe in our modern physical universe. This is a fact we tend not to dwell upon because of the disconcerting implications. People in the past strongly believed in the truth of their universes, and because they were so greatly mistaken, might not we be a little mistaken also, and if a little, why not a lot? We dismiss the thought on the grounds that our knowledge is greatly superior. But knowledge guarantees neither wisdom nor truth, and the thought persists. The early people of a hundred thousand years ago had brains as large as our own, thirty thousand years ago some had brains even larger, suggesting that the universes in which they lived, or thought they lived, were possibly as richly elaborate as those of more recent societies.

If the past is a guide to the future, our modern beliefs might also be greatly mistaken, and one day a new universe might arise, grander than our present model. Those living in the future will look back in history and see our universe as out-of-date like all the rest. In a hundred thousand years they might wonder what we were doing, or not doing, with our large brains.

*   *   *

Thomas Huxley wrote in 1869 for the first issue of the now widely read science journal *Nature*, "It seemed to me that no more fitting preface could be put before a Journal, which aims to mirror the progress of that fashioning by Nature of a picture of herself in the mind of man, which we call the progress of Science." I paraphrase Huxley by saying that the Universe, through us, fashions pictures of itself that we call universes. They are not fancy-free inventions "begot of nothing but vain fantasy," and we are not dreamy playwrights spinning "insubstantial pageants" and "baseless fabrics out of thin air." Each universe is but one of the numberless realities of the Universe.

George Berkeley, an Irish philosopher and bishop in the early eighteenth century, argued that only our mental experiences are real, minds and God alone exist, and the external world is an illusion emanating from God. James Boswell in his biography of Samuel Johnson wrote, "We stood talking for some time together of Bishop Berkeley's ingenious sophistry to prove the non-existence of matter....I shall always remember the alacrity with which Johnson answered, striking his foot with mighty force against a large stone, till he rebounded from it – 'I refute it thus'." Few persons would disagree with Johnson's impressive demonstration of the concreteness of the external world. Although the facts of the external world are certainly more than mere ideas, yet they are rarely as solid and secure as they seem. "Where," asks Morris Kline in *Mathematics in Western Culture*, "is the good, old-fashioned solid matter that obeys precise, compelling mathematical laws? The stone that Dr. Johnson once kicked to demonstrate the reality of matter has become dissipated in a diffuse distribution of mathematical probabilities." The facts are far fewer, the ideas dressing the facts far more, than we normally suppose.

Arthur Eddington, a scientist who leaned toward philosophy and wrote fascinating books that lured the youth of my time into physics, once said, "We have found a strange footprint on the shores of the unknown. We have devised profound theories, one after another, to account for its origin. At last we have succeeded in reconstructing the creature that made the footprint. And lo! it is our own.... The mind has but recovered from nature what the mind put into nature." Eddington took the view that our minds shape our knowledge of nature. This makes sense if *nature* has two meanings: universe and Universe. Our minds shape our knowledge of the Universe in the form of a universe.

A Leibnizian view that has some appeal, despite its vagueness, is that the Universe is an all-encompassing Mind (whatever that means) that contains our individual minds, and the universes are our minds perceiving and seeking to understand the Universe. But this tentative view is no more than a model, barely deserving the name universe.

* * *

The *Masks of the Universe* divides into three parts. Chapters in the first part cover some universes of the past: the magic, mythic, geometric, medieval, infinite, and mechanistic universes. These chapters are brief case studies of the cosmic belief systems of earlier societies, chosen for their historical interest and contribution to modern cosmology.

I start with a speculative account of the magic universe that I imagine arose hundreds of thousands of years ago when *Homo sapiens* had acquired advanced linguistic skills. The magic universe, which began as an animistic world actuated by psychic elements, developed into a living world, vibrant with ambient spirits motivated by thoughts and emotions mirroring the thoughts and emotions of human beings. Mankind's inner world was projected into the outer world. Hosts of spirits of every kind pervaded the magic universe and conformed to codes of behavior resembling the primitive social codes regulating human behavior.

The word "magic," as used here, does not mean the miraculous. It denotes whatever in the external world manifests human characteristics and mimics human behavior, such as apparitions, angels, ghosts, fairies, and the like. In the magic universe, the inner mental world is projected into the outer world, and humanlike motives and impulses serve as the activating agents. Perhaps nobody in the last ten or so thousand years has known what it is like actually to live fully immersed in the magic universe.

Across the span of hundreds of millennia the magic universe evolved into a constellation of magicomythic universes. The ambient spirits of the magic universe were swept up into the empires of potent spirit beings who personified the phenomena of the external world. Many of the multivalent magicomythic universes survived until recent times in out-of-the-way places of the globe.

The mythic universe (mythic because its elements now fail to fit naturally into the modern physical universe) arose less than twenty thousand years ago. It was an enlarged universe ruled by powerful gods who controlled and created all that existed. This new and unified world view reached an advanced stage in the delta civilizations of Mesopotamia, Egypt, and India, and attained its highest forms in the Zoroastrian and medieval universes.

The mythic universe was purchased at a high price. The world of matter – of clouds, rocks, plants, and animals – became spiritless and dead. In an enlarged and transfigured world, riven by the dualities of good and evil, soul and flesh, fate and free will, the timeless tales of the mythic universe tell of the tyranny of divine kingship, of incessant sacred wars commissioned by gods, of appeasement of the gods by human sacrifice, and of the massacre and enslavement of people worshipping other gods.

In the Hellenic world of classical antiquity we see the rise of scientific inquiry and its rejection of the gods as the proper agents of explication. Out of the Ionian, Pythagorean, and Eleatic schools emerges the influential Aristotelian, Epicurean, and Stoic world systems.

The medieval universe – incorporating Zoroastrian, Hebraic, and Aristotelian elements – arose in the high Middle Ages. This magisterial universe, dominating the historical skyline, was surely the most satisfying world system ever devised by the human mind. Here was an age of scholarship and high adventure in which social and technological revolutions culminated in a style of life unique in history and laid the foundation of modern Western society that has spread worldwide.

Scholars in the high and late Middle Ages formulated notions that opened the way for the development of the Cartesian and Newtonian universes. These world systems, particularly the Newtonian system, rose to eminence in the Age of Reason in the eighteenth century (the century of progress), flourished in the Victorian era in the nineteenth century (the century of evolution), and ushered in the physical universe of the twentieth century that overturned the mythic world of dead matter.

Chapters in the second part of the book deal with the physical universe. I discuss those aspects on which our ideas have changed and are still changing. My intention is to stress what seems most interesting, and to weave into the narrative strands from earlier themes. Beneath the surface of the physical universe lie forms of magic more bewildering than ever before. Science reawakens the dead matter of the mythic universe with an inlay of vibrant activity, and the physical universe is now akin in some ways to the old magic universe. But the coruscating agents of explication dance more brilliantly and intricately than ever before. Much of modern science consists of magic disciplined by a calculus of mythic laws.

In the third part I alight on miscellaneous topics of cosmological interest. I start with the witch universe that arose in the late Middle Ages and terrorized the Renaissance. It serves as a pathological case study of a mad universe. It illustrates a basic point that all universes are verified in accordance with their own rational principles. I then turn to other topics such as containment, consciousness, and learned ignorance.

Cosmology plucks fruit from all branches of knowledge. Wonderful and strange are "the universes that drift like bubbles in the foam upon the River of Time," wrote Arthur C. Clarke in the *Wall of Darkness*. The universes, wonderful and strange, reveal mythic and mechanistic vistas, all constrained in scope by their own criteria distinguishing what is real from the unreal, what is true from the untrue.

*　　*　　*

One important issue concerns the Universe and God. Both are unknown and unknowable in any absolute sense, both are fundamentally inconceivable, and both are all-inclusive. Is it therefore possible they are one and the same thing, and the distinction that we attribute lies only in the models (the masks of God and the masks of the Universe) that we create? I discuss this in Chapter 18, "The Cloud of Unknowing".

From history we learn that the fate of every belief is eventual disbelief. Some thinkers have therefore turned to skepticism and denied all truth. There is one belief, however, that must always endure: belief in a reality veiled in mystery and beyond comprehension. The mystic who wrote *The Cloud of Unknowing* in the fourteenth century came to the conclusion that ultimate reality lies beyond understanding, and was saved from skepticism by reverence of the mystery of existence. The cloud of unknowing is the Universe, and the many universes are our visions of the Universe.

The Universe lies beyond the reach of human comprehension; whereas the universes, which we believe we live in, are comprehensible and rational by their own standards. By distinguishing between the Universe and universes we gain insight into the basic difference between mind and brain, between free will and determinism. The mind with its consciousness and free will, having no natural place in our comprehensible and rational universes, belongs to the Universe.

Part I Worlds in the Making

lived, is lost forever, and all our reconstructions are possibly in error.

2 The Magic Universe

"History is only a pack of tricks we play on the dead," said Voltaire. By scanning history, peering into prehistory, we seek the ancestral incunabula. With meddlesome curiosity we turn over stones, dig up bones, and expect the dead of long ago to forgive the tricks we play.

At least we have learned not to portray early human beings as shambling Nibelungs, or as Hobbesian ogres, "solitary, poor, nasty, brutish, and short." Doubtless the forgotten people of the distant past were thoughtful beings, with a spring in their stride and light in their eyes, who ornamented their bodies, bedecked their dead with flowers, danced, sang, laughed, cried, and, like us, had their joys and sorrows. They lacked our knowledge, yet had instead their own, perhaps more than we can ever realize.

Little is known of the early people who lived hundreds of thousands of years ago. Their lifestyle was certainly primitive by our standards and even by the standards of the African Bushmen and Australian Aborigines. Other than a miscellany of skulls and skeletal remains, tool kits, artifacts, and evidence of diet, we have precious little information on how the early people lived, and none whatever on how they thought. But we know they had brains as large as ours and we may safely assume that their brains, like ours, were fully functional. The universe in which the early people lived, or thought they lived, is lost forever, and all our reconstructions are possibly in error. My guess is the following.

*   *   *

Imagine a nomadic group of hairless and thin-skinned striding primates, encumbered with juveniles who take a decade to reach maturity and elders who need special care. This picture of early people wandering on savannas, along seashores, and through woodland forests prompts us to wonder how they could survive when the animals around them were fleet-footed, protected by fur, and armed with sharp claws, horns, long teeth, and tusks.

True, in their skillful hands the crafts of bone carving and stone chipping had developed into an industry of toolmaking (and let us not overlook furriery, pottery, cookery and other crafts). "Man is a toolmaking animal," said Benjamin Franklin. Tools made possible the weaponry that compensated for a defenseless physique. But we go too far when we credit toolmaking with the breakthrough to large brains. The production of carrying bags (one of the greatest inventions), the control of fire (half a million years ago), and the skills of tool using and toolmaking (as old as *Homo sapiens*) were surely effects and not causes of the breakthrough to large brains.

Our picture of a group of primates equipped with carrying bags, fire, tools, and weapons is incomplete. It omits the supreme fact that they are chattering together. The breakthrough to large brains had started when human beings first began to speak. Language organized and unified social groups that were able to live and rove in unsheltered environments.

Three million years ago the *Australopithecus* hominids of South Africa had a cranial capacity of 400 to 500 milliliters, already larger than that of chimpanzees; a million years later *Homo habilis* had a brain volume of 600 to 700 milliliters; the rate of increase was rapid, and a million years ago the brain size of *Homo erectus* had increased to between 900 to 1100 milliliters; modern human beings soon emerged with an average cranial capacity of 1450 milliliters. (Curiously, for reasons unknown, the size of the human brain has been decreasing over the last thirty or so thousand years.) The principal differences between human beings and apes are brain size and language. We may reliably suppose that the cranial capacity of fossil skulls serves as an indicator of hominid intelligence and the development of mental processes associated with language.

Apes communicate with sounds and gestures, and their signals to one another enable them to live as groups in sequestered environments. But the structured articulations of language are far more than just a repertory of sounds and gestures. "Language is a...noninstinctive method of communicating ideas, emotions, and desires by means of a system of voluntarily produced symbols. These symbols are, in the first instance, auditory and are produced by the so-called 'organs of speech'," wrote Edward Sapir, a pioneer of modern linguistics, in his popular book *Language*.

We have a picture of a tightly knit group of jabbering individuals who share their thoughts and feelings. They live on a mixed diet, hunting and gathering, and it is a fair complaint, remarks William Howells in *Evolution of the Genus Homo*, "that man the hunter has been extolled at the expense of woman the gatherer." Men and women, then as now, had equivalent opportunities for the exercise of intelligence and courage. Much to our surprise, the early people did not live in constant fear of a hostile world. They consulted together, formulated plans, acted on command as a unit, referred to a cultural memory of effective strategies, and employed devastating tactics of alternating offense and defense. With language was forged the mightiest weapon on Earth. Men and women are talking animals.

Language raised intelligence to higher and ever higher levels, and articulate thoughts interlaced facts in a widening expanse of memory. Greater intelligence made possible more intricately structured forms of speech. And intelligence was naturally selected, for whoever could not find the apposite words, comprehend and obey the voice of command, recall the effective strategy, or respond with the efficient tactic, had much less chance of surviving. Behold! Men and women are heroic animals, for the early people trod a perilous path of awesome challenge. Perhaps many hominid species started and failed, perhaps some retreated back into sequestered and less-perilous worlds. Chimpanzees, it has been suggested, are perhaps dehumanized hominids who withdrew from the challenge.

We lack a generally accepted method of measuring intelligence. Let us not forget entirely, however, that nature once had her own, perhaps still has, and dispensed judgment in her forthright fashion. Candidates with low scores were eliminated and modern men and women are the prize-winning products of that hard school.

Children take a long time to reach physical maturity, and human beings have evolved that way because many years are needed to learn the language and cultural heritage. This alone indicates how great was the knowledge our remote ancestors handed on to their offspring. In the hunting and gathering groups, the young were taught the language and initiated into the tribal laws and cosmic truths, and the old were cherished as wise leaders and guardians of the cultural memory. Social groups indifferent in the care of their young and old did not survive for long. The lifestyles of the Aboriginals of Australia, the Shoshones of North America, the Pygmies of the Congo Valley, and the Bushmen of the Kalahari Desert offer clues concerning the lifestyles of the early people, but the clues are slender and possibly misleading.

*   *   *

Anthropologists have speculated on how the people of long ago might have viewed their world. In *Before Philosophy: The Intellectual Adventure of Ancient Man*, Henri and H. A. Groenewegen Frankfort, John Wilson, and Thorkild Jacobsen suggest that the world appeared to primitive humans "as neither inanimate nor empty but redundant with life." Everything was living:

> Life had individuality, in man and beast and plant, and in every phenomenon which confronts him – the thunderclap, the sudden shadow, the eerie and unknown clearing in the wood, the stone which suddenly hurts him when he stumbles while on a hunting trip. Any phenomenon may at any time face him, not as "It," but as "Thou." In this confrontation, "Thou" is not contemplated with intellectual detachment; it is experienced as life confronting life, involving every faculty of man in a reciprocal relationship.

The early people lived in a world animated by life. Their comprehension of the world consisted of knowing that everything was alive. The difference between being animate and inanimate was no more than the difference between being awake and asleep. In the opening act, possibly thousands of millennia before the present, the world was little more than an animation in which things had their identifying names and distinguishing patterns of behavior. The inner psychic states of the animata had no distinction from their outer physical forms.

In time, the early people discovered the depths of personality and enlarged their world by conceding to one another an inner mentality of thoughts and feelings expressed in a wealth of linguistic terms. Each person knew that the motives and emotions of other members of the social group were similar to his or her own. Greater intimacy in family and social living followed. Probably at this stage man the hunter and woman the gatherer became mutually supporting within a stable family unit. Inevitably, the projection of the inner self into other persons widened to include beasts and plants, and everything else that called for attention. At last, we stand at the threshold of the magic universe.

*   *   *

Human desires and impulses animated all things and the magic universe was alive in every conceivable sense. The external world mirrored the human mind. It was a looking-glass universe capable of explaining the entire range of human experiences. The evolving human mind, continually strained to its limits, was reflected in the progressive enrichment of the external world.

Out of a total population of several million hominids, only a few social groups, each of a few hundred members, first crossed the threshold into the magic universe. Their newfound imaginative power gave them a superior ability to survive.

The word magic is widely and loosely used in many contexts. Here I have honed it down to mean little more than the human mind made manifest in the external world. If you believe in angels, fairies, demons, ghosts, vampires, the evil eye, and other anthropopsychic agents activating the external world, then you live in a sort of magic or thaumaturgic universe. But nothing like the world of the early people, for their world was totally real and not just a virtual world of superstitious fantasy.

At some stage, still long ago, many of the activating psychic entities of the magic universe attained a kind of independent existence. The inner psychic being became detached from or only tenuously connected to its physical body. Many of these psychic beings – or spirits – endured after the dissolution of the physical body.

The animated world deepened into an animistic world that everywhere was densely populated with embodied and disembodied spirits. Animism is the belief system that all material things have their indwelling spirits. Perhaps the early people supposed that life never died and the inner self gained freedom, as in dreams, and became a spirit. No doubt the early people had a different view of time, and events of the past, present, and future coexisted, and nothing died but transformed from a corporeal to an incorporeal state. Perhaps, by growing aware of an inner mentality as distinct from the outer physical body, the early people automatically attributed this dichotomy of the inner and outer self to everything else, and spirits became the reified mentalities of the external world.

Through deeper understanding the early people gained greater control of the phenomena of their world. Language expanded in scope to encompass the concepts of detached and diffuse spirits. Rivers, lakes, mountains, valleys, and clearings in woods acquired their own ambient spirits, and diffuse nature spirits invested the earth, mountains, sky, wind, water, and fire. The magic universe, pulsating with spirit activity of every kind, reflected and magnified the emotions and thoughts of human beings. A veneer of physical forms overlaid a vibrant world of benevolent, indifferent, and malignant spirits that resonated with the inner world of each person and amplified all mental experience.

A magic universe each day was awakened by the Sun spirit and at night mourned by the Moon spirit. It was a universe of starlike campfires stretching across the night sky, of chromatic sky spirits manifesting in rainbows, sunsets, and northern lights, of mighty earth spirits rumbling beneath the ground and spewing forth from volcanoes, and of flittering little folk dwelling in secret places and stealing lost children. It was a universe haunted by the dead forever calling. Words cannot recall nor the mind recapture the intense vividity of its imagery. On stormy nights the trees awoke, swaying their contorted branches, conspiring in sibilant voices, creating abject terror among huddled people and their familiar spirits. The sudden noise, the fallen tree, the shaft of light piercing the forest gloom, the rising river, the lowering sky, the hurtful stone, and each incident of every day was the natural outcome of incarnate spirits pursuing their personal interests. It was a numinous world of the kind fleetingly glimpsed by children in spine-tingling fairy tales.

*   *   *

The magic universe was fully rational in accordance with its principles. We must put aside the tales that primitive people could predict nothing because of spirit capriciousness. Humanfolk and their spiritfolk were no more capricious in behavior than we are today.

Spirit behavior reflected human behavior, and human beings predicted the acts of spirits to the extent they predicted the acts of one another. A rebellious person might be coaxed by soothing words, loved by concerned kinfolk, shown in what way he or she stood to gain by conforming, shamed by indignation, and occasionally coerced by dire threats. Similarly, the spirits could be coaxed, loved, bribed, shamed, and coerced. By offering gifts and performing pleasing tasks, the people influenced the spirits in the same way and to the extent they influenced one another. Thus the aid of benign spirits was enlisted and the harm of malign spirits averted.

By reading the signs, the early people gained control over their universe and predicted many of its events. The lowering sky gave warning of the imminence of storm spirits, and the forewarned people took shelter. A child while running for cover with its mother might trip on a stone, and after the mother had scolded the hurtful stone the child was never again tripped by the same stone. Always the spirits displayed signs that made clear their moods and intentions, and the people read the signs and acted accordingly. By constant dialogue and by coaxing, loving, bribing, shaming, and coercing the spirits the early people were able to influence and control their world.

The magic universe consisted literally of life confronting life. What seems to us an ineffectual cosmology, on the contrary, seemed to the early people fully effectual. They probably had more understanding and control of their world than we individually have of our world. Few people today understand how internal combustion engines, jet engines, telephones, computers, and the internet work, how airplanes fly, or how to repair television sets. Yet these are now the commonest things around us. The early people not only lived in a comprehensible world, but also knew how to influence and control it, which is more than can be said of most of us today. I am inclined to think that of all known universes, the magic universe was in its own terms the most rational and lucid, and all subsequent cosmological developments have been purchased at the cost of added mystery and perplexity.

* * *

"Possessed, pervaded, and crowded with spiritual beings," said the Victorian anthropologist Edward Tylor, referring to the world of primitive people. In his *Primitive Culture* of 1871 he proposed the theory of animism and conjectured that animism was invented by "ancient savage philosophers." Theories of how the early people thought are no more than guesswork, and if animism is the correct theory, as I have assumed, it seems unlikely that it originated as a philosophical invention. More likely, as intelligence advanced, the animation of objects evolved naturally into the animism of objects ruled by spirits.

The Sorcerer. A paleolithic cave painting from the French Pyrenees.

"I shall invite my readers," wrote Branislaw Malinowski, "to step outside the closed study of the theorist into the open air of the anthropological field." We buy our tickets and accompany Malinowski to the Trobriand Islands of Melanesia. There, on these islands, as described in *Magic, Science and Religion*, we find *mana* (a generalized spirit), totemism, shamanism, sorcery, cults of vegetation and fertility, fetishes and charms. The Trobriand Islanders work in their gardens and fish from their canoes, drawing on a large body of empirical knowledge, and their beliefs in the supernatural are inconspicuous except for the shaman's ritual of occasionally blessing the gardens and canoes. Religion takes center stage in rites of passage and ceremonies, and particularly in invocations of powerful spirits when preparing for long voyages, fishing in hazardous waters, or taking to arms in time of war. Supernatural beliefs hang in the background like a tapestry weaving together the threads of mortal and immortal life, and social customs and traditions stand prominently in the foreground regulating the affairs of everyday life. This is certainly not the magic universe. It looks not unlike many universes of the recent past and present.

Nowhere in the anthropological field can primitive animism be found. In fact, animism fails to explain the sophisticated belief-systems of recent and present-day "primitive" societies. The word primitive, denoting what is earliest or among the first, confers a deceptive aura of simplicity. Call a thing primitive and the battle of explaining it is half won. Often we label out-of-the-way people primitive when their lifestyles and belief-systems are other than our own. The word is a misnomer that leads us much astray.

One might justly wonder whether in historical times any truly primitive society has existed. The societies familiar to us look much too sophisticated to be dubbed primitive. Their languages and beliefs are as rich and complex as those of non-primitive societies. The assumption that our society has evolved from primitive societies similar to those now existing is equivalent to assuming that we have evolved from apes similar to those now living. We and contemporary apes have diverged over great periods of time from early primates, and similarly, the societies covering the globe have diverged over great periods of time from earlier societies. The magic universe no longer exists.

*   *   *

The magic universe evolved and lost its simplicity. Hitherto, the lifestyles of spiritfolk had reflected little more than the lifestyles of humanfolk. One side mirrored the other. As human societies evolved, so did the spirit societies, and one side continued to mirror the other.

But in time the mirror began to distort and magnify the spirit images. With more knowledge came a growing awareness of the vastness and complexity of nature and a realization that the beings responsible for activating the world were greatly superior to humanfolk and ordinary spiritfolk. Step by step the magic universe evolved into a magicomythic universe. On one side of the mirror stood human beings, on the other side towered superspirits – veritable godlings – who orchestrated the large-scale phenomena of the world and exercised abilities never granted to human and spirit folk.

The little spirits who once had activated everything in haphazard fashion, or so it now seemed, who needed to be constantly watched and cajoled into compliance, were absorbed into the empires of the godlings. Those that managed to survive vanished into secret places.

Ceremonial worship of the godlings replaced the old spontaneous dialogue with spiritfolk. Incantations appeased mighty and fearsome spirits. Invocations and sacrifices sustained the rhythm of the seasons and guaranteed maintenance of food supplies. To hunt and kill required permission not from the animal itself, as in earlier ages, but from the spirit of its species, obtained through the medium of the totemic shaman. This kind of magicomythic universe, controlled by superhuman beings and nature spirits, is what we find in the anthropological accounts of "primitive" societies.

The timid spirits of the magic universe had never shown much interest in distant places. Not so the godlings of the magicomythic universe who ruled far and wide. Each society believed in its central importance in the scheme of things and in the superiority of its godlings. Rival godlings, intolerant of one another, drove their social groups into open conflict.

The many worlds of the magicomythic universe collided and erupted in turmoil. Only when overwhelmed by conquest would a social group accept the godlings of the victorious group. Those groups unequal to the challenge either melted away or fled to the security of outlandish regions. Those magicomythic worlds with the mightiest spirits evolved into the mythic universe of advanced civilizations.

*   *   *

Thomas Hobbes, a sixteenth-century English philosopher, argued in *Leviathan* that material laws are fully capable of explaining the characteristics of human behavior. Chemistry, biology, the cognitive sciences, and sociology have confirmed much of Hobbes's argument. Furthermore, he argued that ethics must be freed from its bondage to religion and grounded on rational premises. In this also, according to anthropology, it seems that he was mostly right.

Societies display a remarkable diversity of religious beliefs and an equally remarkable uniformity of moral codes of behavior. In "Religion and Morality" (*Encyclopedia of Philosophy*), Nowell-Smith discusses the religious diversity and ethical uniformity in various societies and draws the conclusion that moral codes are not of religious origin. Contrary to widespread thinking that without religion there can be no morals, the anthropological evidence indicates that moral codes are of greater antiquity than current religious beliefs. Murray Islanders teach their children the importance of truthfulness, obedience, respectfulness, and kindness to kinfolk. Uncivil acts, such as shirking duty, abusive language, and borrowing without permission, are forbidden, and Nowell-Smith adds, "Similar lists of rules can be cited from many primitive tribes, and the lists might have come from a present-day pulpit or classroom."

Moral codes and rules of conduct have probably existed as long as human beings have lived together in social groups. Hominids for millions of years and human beings for hundreds of thousands of years have lived in social groups, and the protocols of mutual support that preserve a social group were thrashed out and sifted by natural selection. Groups composed of liars, thieves, rapists, and murderers had no more chance of surviving than the proverbial snowflake on a summer's day. The codes that consciously and unconsciously regulate individual behavior were once indispensable for survival of the social group, and the social groups weakened by dissident and immoral behavior were eliminated by the iron law of natural selection.

In civilized societies, religious institutions preach and political institutions legislate variations of the old moral codes. They also invent the exemptions and additions. Priests claim that the divine cause justifies every means, politicians claim that flexibility is the highest principle. Specious arguments that override moral obligations can always be found, and fortunately for the human race these arguments are less durable than the primitive moral imperatives.

*   *   *

We cannot recreate the magic universe and recapture its experiences. No social group in the last thirty or more thousand years has known what it was like to live in the age of magic. Not impossibly, primitive human beings lived in a universe more emotionally fulfilling and intellectually demanding than the universes of most societies in recent times.

3   The Mythic Universe

The changeover from the magic universe to the mythic universe never reached completion in Australasia and other isolated lands secure from assault. The populations in these lands survived until recent times snug in their halfway magicomythic worlds. Elsewhere, the globe was in uproar with the rise of the mythic universe.

Climate changes and cultural conflicts stirred the swirl of tribal movements. Food hunters and food gatherers turned to herding and farming, and farming communities emerged between ten and twenty thousand years ago in the Middle East, India, China, Africa, Europe, and later in Mesoamerica. Tribes multiplied, merged and became nations. Powerful ruling families attained royal status, and professional priests interpreted the will of the gods. The arts burgeoned into professions and the crafts into industries. Irrigation systems connected rivers to farmlands, and large works such as Stonehenge in Britain and the pyramids in Egypt marked the rise of engineering. Trade flourished over great distances, as between the cities of Sumer and Akkadia in Mesopotamia and the far cities of Mohenjo Daro and Harappa in India.

*   *   *

The mythic universe was well under way more than six thousand years ago with the rise of the great gods in the delta civilizations of the Nile, Euphrates–Tigris, and Indus. "Thou art the Sole One who made all that is, the One and Only who made what existeth," chanted the Egyptian priests of the New Kingdom in adoration of Amun the god of Thebes. In the new cosmology all things were created and controlled by all-powerful gods who dwelt in far-away places.

In the magic universe nature throbbed with spirit life; at the other extreme, in the new mythic universe, all this pulsating liveliness was withdrawn from the natural world and given to the gods. The world, squeezed dry of the spirit of life, became totally lifeless. Dialogue with spiritfolk, who once dwelt everywhere, transformed into worship of gods and goddesses who dwelt in cosmocryptic realms high in the sky or deep underground.

The mythic universe was more than just the magicomythic worlds outfitted with greater gods. Beasts, plants, and everything else still displayed the same outward forms but the inner spirit had gone. Trees no longer suffered pain when felled and pleas for their forgiveness were unnecessary; the fire no longer was nurtured with loving care in fair return for its warmth and light; no need to beg permission of the wood spirit before entering the forest, the water spirit before fording the river, the bison spirit before engaging in the hunt. Beasts were kept in flocks and herds to facilitate their exploitation and were slaughtered without apologetic ceremony. All was done by permission of the gods, granted in return for ritual and sacrifice.

In adoration, men and women worshipped the gods of the mythic universe, and in return the gods endowed the world with order and design. Through the machinations of these beings it was at last possible for men and women to comprehend the grand design of the created world, and by sacrifice and prayer they could influence and predict events as never before.

A worn-out magic universe was traded in for a brand-new mythic universe, and though much was gained by the transaction, the price paid was exorbitant. It was not assumed but known that the natural world was dead and devoid of spirit. The evidence of one's senses gave direct proof that the world consisted of spiritless matter. When a person kicked a rock or cut down a tree that person did not injure the gods, who were elsewhere and could not care less. Plain for all to see was the difference between living things and dead matter. Foremost among living things, other than the gods and oneself, came one's kinfolk and members of one's social group who worshipped the true gods. Everything else was bereft of spirit. Much too easily in

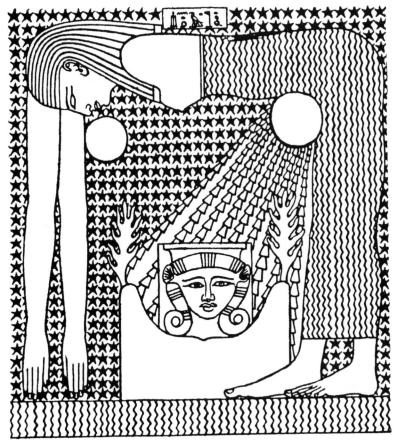

Nut the Egyptian sky goddess gives birth to the Sun whose rays fall on Hathor, the god of life and love. The Earth below is Geb, the brother of Nut.

the mythic universe animals were denied emotions and a capacity for feeling pain. Much too easily people of other races, members of other societies, and worshippers of other gods were denied human status and were massacred, sacrificed, and enslaved.

We see the mythic universe as a dark material world ruled above by shining gods. The deadness, vileness of matter stand out as its distinctive feature. Little wonder that in the Upanishad scriptures, Buddhist teachings, and Gnostic and Neoplatonic theologies we find an abhorrence of the dead material world, its total rejection, and the advocation of world-denying asceticism.

*   *   *

Everything in the old magic universe behaved freely and independently. Everything in the new mythic universe behaved obediently as if jerked by strings in the hands of heavenly puppeteers. All that was free was evil, and all that was virtuous was slave to the gods.

Those societies left clinging to the primitive magic worlds had not the ghost of a chance. They disappeared, annihilated by the organized vigor of the societies of the mythic universe. Plausibly, the great migrations of tens of thousands of years ago into outlandish places consisted of the magic tribes fleeing the rising power of the mythic societies. Wherever a mythic universe brushed against a primitive world that world vanished. In the last millennium hundreds of magicomythic societies have perished. Thus the Tasmanian aboriginals have gone, eliminated by massacre, disease, and the apathy induced with takeover by an alien incomprehensible culture.

History unrolls in the age of gods as a chronicle of tyranny, warfare, human sacrifice and slavery, disclosing the uttermost depths of human misery. This vast expanse of wretched turmoil and the loss of veneration for the natural world lie on the debit side. On the credit side lie grandiose cosmic concepts, otherworldly visions of harmony and law, and lofty abstractions that organize and unify the universe. While gazing over the familiar historical scene, let us remember: not the gods who caused untold suffering, but human beings who created the gods, and thereby organized and directed the immense energies of the human mind.

*   *   *

Myths lack a general definition. Social anthropologists studying Amerindian mythology do not share the views of scholars steeped in Greco-Roman classic literature, and neither are in tune with students of comparative religion. For my purpose the simplest definition suffices. A myth is anything lifted out of another universe that fails to fit naturally into one's own universe. What fits naturally into the modern physical universe, such as Babylonian arithmetic and Euclidean geometry, is prescient and not mythic; what fails to fit naturally, such as Saint Anselm's empyrean and Dante's hell in *The Divine Comedy*, is outmoded and mythic. Although now incredible to us, each myth was once credible in its original cosmic setting.

Mythology is the alchemy of myths. When societies collide, intermingling their cultures, their myths react to form new mythic compounds. The warfare of gods and the victory of right over wrong illustrate symbolically the warfare of nations and the victory of one nation over another. Barbara Sproul in her *Primal Myths* describes how

> ... these myths tell of great battles between the old, degenerate
> gods of the conquered people and the young, energetic gods of the
> conquerors. The earliest Creed of the Celts and the Maori
> Cosmologies both tell of the successful rebellion of divine sons
> against their primordial parents and reflect the triumph of new
> cultures over indigenous ones.

In Mesopotamian myths the old Sumerian female god Tiamat is defeated by Marduk, the warring deity of the victorious early Babylonians, and in Hesiod's *Theogony* the male sky god of the invading Indo-European-speaking people overthrows the female Earth god of the Pelasgians and Cretans.

The creation of the world formed an integral part of the mythic universe. The oldest creation myths, according to Joseph Campbell in *Primitive Mythology*, drew on the generative function of the female body as their central theme, and the created world was a polarization of male and female elements. Neolithic cosmology, and presumably earlier cosmologies, made little or no distinction between the creation of the organic and the inorganic realms, and all animate and inanimate things were born together from a cosmic womb. The creative act involved all of nature, and the newborn world emerged as an organic whole. In the myths of later ages, the creation of the living and nonliving tended to be distinct events: creation was a sequential process, often a twofold act, in which the living and nonliving worlds were created separately, either one or the other coming first.

The 5000-year-old Sumerian epic of creation, *Enuma Elish*, tells that in the beginning, "when heaven above and earth below had not been formed," there existed the primal Apsu – a watery abyss – and the primal female being Tiamat. Apsu and Tiamat begot Anu the sky god who with Tiamat begot Ea the earth god of wisdom. Eventually, six hundred or more gods and goddesses controlled the many realms of existence, and from our matter-of-fact stance they appear to have done little more than squabble incessantly with one another. Following the rise of Babylonia nearly four thousand years ago in the reign of Hammurabi, Ea usurped Apsu and with Tiamat begot the fearsome four-eyed Marduk. Then Marduk overcame Tiamat, divided her into the Upper and Lower Worlds, and usurped numerous gods by appropriating their functions and names. A tripartite universe consisting of Heaven, Earth, and Netherworld emerged in which the wheeling stars and wandering planets disclosed the intentions of the gods.

The earliest Greek myths, recounted in the *Theogony* (History of the Gods) by Hesiod in the eighth century B.C., declared that in the beginning there were four primal beings: first came Chaos the Limitless Void, then Gaea the Earth, Tartarus the Lower World, and Eros the Spirit of Love. These four beings generated arrays of gods who personified all aspects of the universe. The raping of Gaea by the sky-god Uranus gave birth to the Titans, the first rulers on Earth. Uranus and Gaea begot the gods Cronus (ancestor to Zeus), Prometheus (Forethought), and Epimetheus (Afterthought). From out of dead clay Prometheus modeled human bodies in the likeness of gods and breathed into them the spirit of life. Hesiod in *Works and Days* tells how the earliest people were a golden race who lived free of evil and harsh toil. The earthly paradise ended when Pandora, the wife of Epimetheus, committed the original sin of releasing the evils and diseases that Prometheus had locked away. Then came a silver race that neglected to worship the gods, followed by a warlike bronze race, followed finally by a destructive iron race that still lives.

According to the Norse myths of the *Elda Edda*, out of a "yawning chasm" at the dawn of time arose the Frost Maidens, bringing with them Ymir, the first of the giant gods. Their descendent, the one-eyed Odin, slew Ymir and divided his body into Earth and Sky. An apocalyptic element enters the cosmic tales, and in the *Ragnorak* and the *Götterdämmerung* (Twilight of the Gods) of Norse and Germanic folklore we encounter instances of eschatological myths foretelling the end of the universe. From the beginning the world is doomed and men, women, gods, and goddesses are destined to die in a cosmic cataclysm. The end is foreshadowed by baleful omens, oathbreaking, and titanic warfare among gods and men. Amidst the carnage of Doomsday, the Sun becomes swollen and blood red, and the Earth in the grip of paralyzing winter sinks back into the chasm. Out of the cosmic wreckage arises a new universe of "wondrous beauty" ruled by other and perhaps better gods.

*   *   *

The *Epic of Gilgamesh* comes from the Babylonian records of around five thousand years ago. Gilgamesh, a young Sumerian king of the Uruks, lives a riotous life. In response to complaints from the citizens of Uruk, the gods create Enkidu, mortal and strong, to curb the excesses of Gilgamesh. In time, Gilgamesh and Enkidu become friends and share many adventures. In one adventure they overcome and destroy the Bull of Heaven, and for this impious deed the gods exact retribution and Enkidu dies. The death of Enkidu shocks Gilgamesh, and in his grief he cries out, "How can I rest, how can I be at peace? Despair is in my heart." Born of a mortal father and an immortal goddess, he himself is only half divine and therefore fated to die. Watching the slow corruption of Enkidu's body, he realizes the irrevocable nature of death and at last understands what the denial of immortality means. He rages against his fate, "What my brother is now, that shall I be when dead," and he condemns the gods, "When the gods created mankind, death for mankind they set aside, life in their own hands retaining." Far and wide he journeys seeking to escape his fate. He crosses the Waters of Death to consult with Utnapishtim, the Sumerian archetype of Noah. By surviving the Flood, Utnapishtim and his wife are the only human beings to have been granted immortality. "Because of my brother," declares Gilgamesh to Utnapishtim, "I am now afraid of death. Because of my brother, I stray through the wilderness. His death lies heavy upon me. How can I be silent, how can I rest? He is dust and I shall die and be laid in the earth forever." But he receives no consolation, and learns there is no escape from death.

The *Epic of Gilgamesh* exemplifies in a legendary figure the tragedy of death. The epic is as poignant today as it was five thousand years ago.

*　　*　　*

Endemic warfare came with the mythic universe. Herbert Butterfield in *The Origins of History* states, "it is one of the surprises of history to learn for how long and over how wide an area, war was a sacred thing, and was particularly associated with the action of gods." When in ancient times a monarch went to war, Butterfield writes,

> ... he would feel he was commissioned by the gods to undertake the enterprise. By appeal to the oracle or by various kinds of divinations, he would seek to know the will of the gods, taking action only at their command or when he was sure that he had their favour. It was the god who won the victory, sometimes to the discomfiture of another god.

The gods commissioned the wars, then determined their outcome by various ploys, such as depriving an army of courage, enfeebling it by hunger, or wasting it with disease.

The city-states of Sumer, each of thirty or more thousand citizens, produced skillful works of art, and their crafts and industries supported a high standard of living. The citizens of each city-state, goaded by the inspired dreams of their king and the divinations of their priests, strived to establish the supremacy of their local gods. Wars between the states were waged at the behest of the gods. When one state caused offense by extending its boundaries or by transgressing in some other way, it was the patron deity of the offended state who felt most aggrieved and demanded that the citizens take to arms.

Victory in battle was the just reward for obedient and reverential worship, defeat the punishment for inadequate worship and inattention to details of ritual. There were no permanent armies of trained soldiers, no carefully planned and prepared campaigns, no contrived strategems. Why should there be, when all was in the lap of the gods? The king who relied too much on a large army, or planned ahead too carefully, might lose everything as a punishment for failing to have faith in his patron deity. The best insurance was to promise one's god lavish ceremony and ample sacrifice in return for victory, and in time of war, sacrifice meant human sacrifice. The just fate of all wicked enemies who opposed and angered one's god was death, and the god could be appeased best by making the wicked die painfully. In the sacred wars – every war in the mythic universe was sacred – booty and captives were the property of the gods, the former went to the temple and the latter to the torturer and the fire.

Astonishing as this may seem, remarks Butterfield, European history of the last millennium is no better, perhaps worse, presenting innumerable instances of sacred wars commissioned by angry and jealous gods, of the outcome of wars determined by the gods (victory to the faithful, defeat to the unfaithful), and of the inhumane treatment inflicted by the righteous on all heretics who failed to worship the true gods.

Always men and women have cried out to the gods in time of need. "O God," implored the pious Ashurbanipal, king of the Assyrians in the seventh century B.C., "how long wilt thou deal with me thus?" Well had he served the mighty god Ashur, defeating his enemies, sacrificing in numerous ways vast numbers of male and female captives, piling high the dead and dying to the glory of his Lord, and here he was in old age beset by tribulations and without just reward.

After the death of Ashurbanipal, the Assyrian Empire fell to the Medes and Babylonians, and under the rule of Nebuchadnezzar and his successors the Persian Empire with its more ethical religion rose to power.

The Old Testament tells of a semi-nomadic people racked by misfortunes and ruled by a tribal deity intolerant of all other deities. Led out of bondage in Egypt (about the twelfth century B.C.) by Yahweh, and thereafter resident in deserts on the outskirts of great empires, tossed and turned by the vicissitudes of imperial conquests, the chosen people resolutely clung to their god Yahweh. The power of Yahweh grew in proportion not to the fortunes but to the misfortunes of his people. Great was Yahweh's vengeance against all who opposed and oppressed the Hebrews, and greater still against those who lapsed in their devotions. More than once in the battle songs of the Old Testament we read of Hebrew armies deliberately kept small in order that victory might manifestly be by Yahweh's decree and not the efforts of mortals.

*   *   *

The prophet Zoroaster (or Zarathustra) lived in Persia in the late seventh and early sixth centuries B.C. and founded a religion – Zoroastrianism – of ethical monotheism. This novel version of the mythic universe transformed the old Persian polytheism (akin to the pantheistic cults of Hinduism) and became the influential religion of the Medes and Persians. In the new monotheism, the Lord of Light – Ahura Mazda – created a universe in which goodness (symbolized by light) must ultimately triumph over wickedness (symbolized by darkness). The primitive moral codes of forbearance and mutual support became the essential elements of the new religious life. Zoroaster preached a theology of rewards and punishments in afterlife in which good people ascended to heaven and bad people descended to hell. With Zoroastrianism came a widespread revulsion against human sacrifice.

According to Zoroastrian scripture, history divides into four eras. In the first era, Ahura Mazda creates a universe of light, and foresees in its shadows the inevitability of suffering. In the second era, primeval man and animal exist in a state of glorious freedom before darkness descends and the Evil Spirit destroys all. In the third era, the seed of primeval man and animal gives birth to modern man and animal in which good and evil coexist. The last era commences with the birth of Zoroaster and will culminate in the apocalyptic victory of good over evil.

During their Exile in Babylon, the Jewish people encountered Zoroastrianism and adapted its apocalyptic message and ethical idealism to their own brand of monotheism. Thereafter, as revealed in the Old Testament, the duality of good and evil became the paramount theme, and Satan, hitherto an angelic minion, was promoted to the role of archfiend (the Zoroastrian Angra Mainyu, Lord of Darkness). Zoroastrianism and Persian culture inspired the Wisdom Literature of the Jews, in which goodness, justice, and wisdom were woven in wondrous words into the religious fabric, as in the books of *Job* ("Where wast thou when I laid the foundations of the earth? declare if thou hast understanding"), *Psalms* ("Yea, though I walk through the valley of the shadow of death, I will fear no evil: for thou art with me, thy rod and thy staff they comfort me"), *Proverbs* ("Wisdom is the principal thing; therefore get wisdom; and with all thy getting get understanding"), and the *Song of Solomon* ("Who is she that looketh forth as the morning, fair as the moon, clear as the Sun, and terrible as an army with banners?").

The Magi, a priestly cast of Medes, preached a form of Zoroastrianism that included a liturgy of chanting and a theocracy of angels and demons that still survives. For a thousand years following the fall of Babylon until the time of Saint Augustine of Hippo in the late fourth and early fifth centuries the ancient world was exposed to Zoroastrianism through the popular derivative religions of Mithraism and Manichaeism, and influenced by its infiltration into Greek philosophy, Jewish prophetic literature, and Gnostic and Neoplatonic theologies. Augustine, who molded Western Catholicism, was a Manichaean, and after his conversion he blended Zoroastrian theodicy with Judaic scriptural history. In *The Eternal City* he compared the Heavenly and Earthly Cities and contrasted otherworldliness and the way of grace and salvation with worldliness and the way of evil and damnation. Zorastrianism survives today in India among the Parsee (meaning Persians), whose ancestors emigrated in the seventh century to escape Islamic religious suppression. Cults with ethically inspiring elements, such as Isis (the divine mother and her child), were not uncommon in the ancient world. But the novel concept of a supreme godhead as absolute goodness full of compassion and concern for human life originated in the pastoral milieu of Persia, and its ethical ideals, in common with those of Buddhism in India and Confucianism in China, are found in the Judaic, Christian, and Islamic scriptural records.

* * *

Many religions counter the fear of personal death with belief in immortality beyond the grave. Some, following Zoroastrianism, go further; they level the playing field between rich and poor, the unfortunate and fortunate, good and bad people with a system of otherworldly rewards and punishments. Evildoers in this life are cast down and punished in the hereafter, and their victims are raised up and compensated; the sick and poor who suffer pain and deprivation in this sad life are uplifted and made joyful in life beyond the grave. This powerful theodicy of heavenly justice rectifying earthly injustice nowadays sustains a large fraction of the world's population through the vicissitudes of life.

The old moral codes of mutual support and the rites of birth, initiation, marriage, and death, existed long before their annexation by religion. Over hundreds of millennia of intimate living in small social groups, *Homo sapiens* evolved into a conscientious and cultured animal, sensitive to and concerned with the needs of others. Zoroastrianism for the first time made religion the custodian of ethical principles. Admirable as this theology may seem, with its promise of rewards and punishments in the afterlife, the wholesale assimilation of the primitive social codes of moral behavior has a serious downside. Persons who reject religion because of its archaic mythical beliefs are left without moral imperatives. Disbelievers find themselves condemned as moral outlaws.

*   *   *

Alan Watts in *The Two Hands of God* writes, "This, then, is the paradox that the greater the ethical idealism, the darker the shadow we cast, and that ethical monotheism became, in attitude if not in theory, the world's most startling dualism." With ethical monotheism came the insoluble paradox of evil. How could a beneficent supreme being create a universe that contains evil? Either all-powerful or all-good, but not both. The paradox stands out clearly in the work of Augustine of Hippo: evilness inheres in the cosmic design that paradoxically claims to be wholly good.

By making heaven the carrot and hell the stick we forget the real purpose of virtuous living. It uplifts the life of the individual and strengthens the bonds of society. Good and evil are attributes of human relationships; they are of social not religious origin, and earn their own reward (the enrichment of individual and social life) and their own punishment (the impoverishment of individual and social life).

*   *   *

What is religion? "The conception of gods as superhuman beings endowed with the powers to which man possesses nothing comparable in degree and hardly in kind had been slowly evolved in the course of history," wrote James Frazer in *The Golden Bough*. Frazer traced the origin of religion to a time when the control of the "gigantic machinery of nature" was taken over by the gods. According to Fraser, the gods are indispensable to religion. Against this, one might argue that religion in some form predates the rise of gods and probably is as old as *Homo sapiens*.

Most religions distinguish between the sacred and profane. The magic and mythic universes stand at opposite extremes in a religious spectrum; at one end, in the magic universe, everything in the world is sacred and nothing profane; at the other end, in the mythic universe, everything in the valley of shadows is profane and nothing sacred. Another scheme classifies all religions into three divisions: the prophetic (Confucianism, Jainism, Judaism, Mohammedanism, Protestantism, Zoroastrianism), the sacramental (Roman Catholicism, Orthodox Eastern Church, Hinduism, Shintoism), and the contemplative (Buddhism, Sufism, Taoism). The prophetic religions stress revelation, the sacramental stress ritual, and the contemplative stress mysticism.

Alfred Whitehead, philosopher and mathematician, in *Science and the Modern World,* had this to say about religion: "It is the vision of something that stands beyond, behind, and within . . . yet eludes apprehension; something whose possession is the final goal, and yet is beyond all reach; something that is the ultimate ideal, and the hopeless quest." These uplifting words apply also to some degree to the goals of art, philosophy, and science.

The basic elements of religion are twofold: ideas and emotions. The ideas (as expressed in doctrines, scriptures, creeds, dogmas) weave the threads of mortal life into a theological fabric. The emotions (as experienced in exaltation, ecstasy, adoration, revelation, veneration, trance) enhance individual well-being and strengthen the social bonds. The religious ideas evoke religious emotions and the religious emotions inspire religious ideas. In art, which aims to express the exquisite with the highest skills, a conceptual superstructure is not essential. In philosophy, which aims to elucidate the world of concepts by critical discourse, and in science, which aims to activate the world with harmonies obeying natural laws, an emotional substructure is scarcely essential. Religion is unique; it makes demands on the whole person in concepts and emotions. It has no substitute; it is as old as social living, as old as the human race.

*　　*　　*

Religious emotions are invariable, having much in common in all societies in all ages, whereas religious ideas are variable, changing from society to society and age to age. The emotions that individuals experience are everywhere alike, the associated ideas are everywhere different and serve the purpose of evoking the emotions.

Orthodox religious institutions generally hold the contrary view that their divinely inspired dogmas are of primary importance and the associated emotions incidental and of secondary importance. Religious conflicts, persecutions, and wars are always over differences in the ideas. Each institution is dedicated to the preservation and dissemination of its own cherished ideas. The emotions that lie at the heart of religion are swept into the background and replaced by ritual.

When dogmatists insist on retaining mythic beliefs that conflict with science, they make the mistake of believing that without their old-time mythic faiths they cannot have religion. Religious ideas consistent with contemporary science can always be fashioned to evoke religious emotions. Many people still hold the Thomistic (Thomas Aquinas) view that when conflict occurs between science and religion, it is due to scientific error because biblical teaching is inerrant.

In the mythic universe we see the rise of grandiose concepts of increasing abstraction, unrooted in emotion. I have in mind Cardinal Nicholas of Cusa in the fifteenth century sitting in his cloistered study, meditating deeply on the omnipotence of God and developing ideas on the nature of the universe that anticipated modern ideas in cosmology. Yet at the same time, this devout cleric organized from his study the torment of Jews and persecution of heretics. All persons failing to conform to his beliefs, however sincere and genuine in their emotions, ranked as sinful beings meriting punishment according to the dictates of his religion. Ideas, not emotions, were all that truly mattered.

The general view that religious ideas are primary and emotions secondary has other unfortunate consequences. The rejection by many people of outmoded religious beliefs leaves them thinking they cannot have religious experiences. Even worse, the moral codes appropriated by religion lose their moorings and are set adrift in ethical relativism.

*   *   *

No cosmologist knows exactly what is the Universe and no theologian knows exactly what is God. The difference between God and gods is discussed in Chapter 18 ("The Cloud of Unknowing"). It suffices here to say that the word "god" is used here to denote a model of God in the same way that the word "universe" is used to denote a model of the Universe. The many universes serve as the masks of the Universe and the many gods serve as the masks of God. In "The Cloud of Unknowing" I tentatively equate God and Universe because both have similar attributes: both are all-embracing and inconceivable. Thus, we give back to the world what long ago was taken away with the rise of the gods.

4 The Geometric Universe

Four thousand years ago the Babylonian sky-watchers charted the heavens, divided the sky into constellations of the zodiac, compiled star catalogs, recorded the movements of planets, prepared calendars, and predicted eclipses. Although skilled in the arts of computation, the Babylonians did not theorize on the laws of celestial motion for they were not scientists but priests paying homage to the sky gods of the mythic universe.

Between the seventh and sixth centuries B.C. intellectual activity quickened in many lands. The teachings of Zoroaster in Persia, Gautama the Buddha and Mahavira the Jain in India, and Confucius and Lao-tzu in China gave birth to ethical doctrines and inspired religions of virtuous living. Meanwhile in the Hellenic world an intellectual movement of a different stamp had begun that would also lead to eventful consequences.

The Greek civilization of scattered cities and colonies formed a mosaic of cultures that nurtured an elasticity of mind. Hellenic philosopher–scientists of the sixth century B.C. developed a style of thought radically different from the mystery-mongering of the Babylonian and Egyptian astrologer–priests. The Greeks awoke the dead matter of the mythic universe. They disentangled the sequences of cause and effect in a world of natural happenings. They looked askance at the sacred myths, developed the rudiments of the scientific method, and to this day science inherits their curiosity and incredulity.

It began with the Ionians, descendants of the Mycenaeans, who inherited the Minoan culture of Crete. A thousand years earlier the Minoan civilization, Europe's first civilization, had reached the pinnacle of its splendor. The Minoan language was not Indo-European and its Linear A script remains undeciphered. Evidence suggests the Minoans were a maritime people who lived in unfortified towns and palaces of spacious courts with no large temples. Colorful frescoes of animals, birds, and fish display a spontaneity lacking in the stylized art of Egypt and Mesopotamia.

This lively civilization – Hesiod's silver race that neglected to worship the gods? – expired suddenly, probably because of an enormous volcanic eruption on the island of Thera, and because of the incursion of warlike Mycenaeans. The Mycenaeans, rich in gold earned as mercenaries aiding Egypt in the ejection of the Hyksos, defeated Troy, as narrated in the Homeric epic, and then withdrew in the eleventh century B.C. from the Greek mainland to escape the ravages of invading Dorians. Whatever spiced Minoan life may have descended to the Ionians of the sixth century B.C. and inspired their acuter minds into revolt against the mythic universe.

*　　*　　*

Thales, born late in the seventh century B.C. and the first of the Ionian philosopher–scientists, lived in Miletus on the Turkish Mediterranean coast. Skilled in geometry learned from the Egyptians, he predicted an eclipse of the Sun using astronomy learned from the Babylonians. The world floats in a primordial sea, he said, and is composed of water existing in many forms. Water is the ultimate constituent of all things, for it lives, flows, and permeates the world.

Anaximander of Miletus, a disciple of Thales, said all things are "according to necessity . . . and the assessment of time." No single substance may be regarded as primary, he argued, for the ultimate is indeterminable. The world consists of intermingled opposites – hot and cold, dry and wet, light and dark – and is animated by their interplay. He was the first, said Agathemerus, "who dared to draw the inhabited world on a tablet" (the first to make a map). Anaximander taught that the world alternated between extreme states over long periods of time, and animals, including human beings, had evolved from primitive creatures in the sea.

We know, said Anaximenes, who also lived in Miletus and was a pupil of Anaximander, that air is pervasive and forever restless. Air is the breath of life and therefore the ultimate substance. Air is flame and fire when rarified, cloud and water when condensed, earth and rock when more condensed. Rarified air is hot, condensed air is cold; and we notice how our breath feels cool when forced between pressed lips and yet is warm with the lips apart. Anaximenes said the constellations of stars were fiery rarefactions high in the atmosphere.

Unlike the Egyptians and Babylonians, the Greeks in the sixth century B.C. lacked reliable historical records. The legendary past – when gods performed miracles on Earth – was separated from them by an impenetrable dark age. Hecataeus, born in Miletus while Anaximander and Anaximenes still lived, founded geography and became the first critical historian. "The tales told by the Greeks are many and in my view ridiculous," he wrote. As a young man he had informed the priests of Egypt that the Greeks could trace their ancestry back for as many as ten generations, and even more, to a time on Earth when human beings were still gods. The amused Egyptian priests had shown him the statues of their high priests, arrayed rank after rank, extending back for thousands of years. The astounded Hecataeus thereupon begun a career that made history a subject of disciplined study.

According to Plato in the *Timaeus*, when Solon, a poet and statesman of Athens, visited Egypt in quest of the past, he was told by an old priest, "Solon, Solon! You Hellenes are perpetual children. Such a thing as an old Hellene does not exist." Then, referring possibly to the Minoans, the priest said:

> You have preserved only the memory of one deluge out of a long
> previous series. You are ignorant of the fact that your own country
> was the home of the noblest and the highest human race. You
> yourself and your whole nation can claim this race as your
> ancestors through a fraction of the stock that survived a former
> catastrophe, but you are ignorant of this because for many
> successive generations the survivors lived and died illiterate.

Herodotus, the "father of history" had a similar experience, and after Hecataeus, every Greek student of history spent a semester in Egypt.

The Ionians initiated Greek prose-writing and raised the Hellenic arts to a high level. With unfettered curiosity they peered into the structure of matter, pondered on the nature of time, conjectured on the distinction between the planets and stars, speculated on geological and biological evolution, and developed meteorology and theorized on the physics of storms. They used mechanical analogies from the arts and crafts, and as the poet Berton Brayley said, "Back of the beating hammer... the seeker may find the thought."

* * *

Overshadowing the ancient world stands the enigmatic figure of Pythagoras. Born on the Ionian island of Samos in the early decades of the sixth century B.C., and perhaps a student of either Thales or Anaximander, Pythagoras is reputed to have traveled widely and imbibed knowledge from many lands. In his later years he taught at Croton in the south of Italy and founded a society similar to the Orphic communities then flourishing in Italy and Sicily.

According to Diogenes Laertius, Pythagoras was "the first to call the heavens cosmos and the Earth a sphere." The universe, said Pythagoras, is like a musical instrument, and the celestial spheres – governed by geometric laws – move with musical harmony in circular paths about an unseen central fire. Pythagoras established mathematics as a disciplined study; he formulated theorems with economy and rigor, and developed geometry to the level at which Euclid inherited it. By experimenting with vibrating strings, he discovered the arithmetical relations between harmonious notes, confirming his conviction that beneath the tumult of common occurrences lies the harmony of numbers. The Pythagoreans – followers of Pythagoras – worshipped a universe suffused with arithmetical divinity and believed, like modern theoretical physicists, that truth is revealed by reducing the world to its numerical elements.

The word *philosophy*, meaning "love of wisdom," comes to us from the Pythagoreans, who sought wisdom with passionate enthusiasm. They had little fear of prudence being swept aside by *enthusiasm* (meaning "possessed by the gods"). Bertrand Russell, mathematician and philosopher, commented:

> To those who have reluctantly learnt a little mathematics in school, this may seem strange; but to those who have experienced the intoxicating delight of sudden understanding that mathematics gives from time to time to those who love it, the Pythagorean view will seem completely natural...It might seem that the empirical philosopher is the slave of his material, but that the pure mathematician, like the musician, is a free creator of his world of ordered beauty.

Possibly the Pythagoreans were influenced by the Orphic belief that revelation is the essence of all religious experience. The Orphic creed of the god Dionysus bore little resemblance to the worship of Bacchus (god of festivity) and possibly was yet another echo of the Minoan culture. *Orgy*, an Orphic word meaning "sacrament that purifies," was later corrupted by association with Bacchanalian revels. *Theory* and *theater* share the same root, meaning "to view." *Theory* came into philosophy and science via the Pythagoreans with the Orphic meaning of "passionate contemplation." Albert Einstein, a modern Pythagorean, wrote:

> The most beautiful emotion we can experience is the mystical. It is the sower of all true art and science...To know that what is inscrutable to us really exists, manifesting itself as the highest wisdom and the most radiant beauty, which our dull faculties can comprehend only in their most primitive forms – this knowledge, this feeling is at the center of true religiousness. In this sense, and in this sense only, I belong to the ranks of devoutly religious men.

*   *   *

Anaxagoras, born in Clazomenea near Ephesus about 500 B.C., just before the cities of Asia Minor fell under the rule of the Persian Empire, was among the last of the outstanding Ionian philosopher–scientists. He lived and taught in Athens at the time of Pericles. "Nothing comes into being or perishes but is compounded from or dissolved into things that endure," declared Anaxagoras. Probably he inspired the atomists by proposing that all things are composed of numerous minute portions (called seeds) of an elemental substance. The Moon, said Anaxagoras, shines by reflected light and has mountains on its surface; the stars are fiery bodies so distant that we cannot feel their warmth. He originated the momentous cosmological idea of a universe of unlimited extent in which things everywhere have similar composition and are subject to similar laws. The universe is ruled by the Mind – the Logos – not the surd gods, said Anaxagoras, and for this impiety he was impeached, and though acquitted, he deemed it wise to flee Athens.

Heraclitus of Ephesus lived in the late sixth century B.C. and was an Ionian of a different stamp. Like most Hellenic thinkers at that time he was influenced by Pythagoras. He taught that the Logos – the Word or God – was the basic unifying principle. He is best known for declaring the world "was ever, is now, and ever shall be a living fire," and "all things change and nothing remains at rest," and we "never step into the same river twice." Only change is changeless and wisdom lies in knowing how things change. Heraclitus envisioned a world in which "things come into being and pass away through strife," and anticipated, we may venture to guess, the notion of "survival of the fittest." The Heraclitean system of whirling bodies and swirling fluids never at rest foreshadowed aspects of the Cartesian mechanistic universe of the seventeenth century.

From the Ionians the scene changes to the Eleatics in the Hellenic city of Elea on the southern coast of Italy. Parmenides, a prominent Eleatic philosopher, had little sympathy for the Heraclitean system of perpetual flux. On the contrary, declared Parmenides, nothing truly changes, for all change is mere appearance and an illusion of the deceived senses. Wisdom lies in knowing that beyond the tumult of transient happenings there exists an "invariant sphere of being," a timeless reality that is reached and grasped by reason alone. Parmenides was the first to make the duality of appearance and reality the basic issue.

We see in the modern physical universe echoes of the Parmenidean timeless view of the world. Consider the following. An event is something that happens at a moment in time, such as a lightning flash or a raindrop falling against my window. Events are arrayed at fixed positions in time, and only our movement in time from the past to the future makes them seem to change – to come and go. Our perceptions (the sights and sounds) are limited to the *now* – a window in time – and as the *now* moves in time, the universe unfolds and reveals through our perceptions what previously was unknown. Suppose we were not bound to the moving *now* and could perceive the whole universe throughout all time. There would be no unfolding, no perception of change, for everything would be simultaneously disclosed in a single timeless act. The idea of the *now* moving in spacetime was described by Charles Hinton in the nineteenth century.

But why are our immediate experiences limited to a narrow window that we call *now*? And why does the *now* move through time, thus creating a world that appears in a state of continual change? Neither Parmenides, nor Hinton, nor anyone else has explained transience. Gerald Whitrow says in *The Natural Philosophy of Time*, "we try to explain transience by assuming transience" in some other form. The transience of the perceived world is generally attributed to motion of the *now*, which begs the question by presupposing transience in the form of motion. Some philosophers and scientists, in the Parmenidean tradition, believe that change does not exist in the physical world and our experience of change is psychological or metaphysical. But again, transience is explained by assuming that it exists in another form (as in the River of Time or the Wheel of Time).

The systems proposed by Heraclitus and Parmenides represented extreme views: total action on the one hand and total inaction on the other. Empedocles of Acragas in Sicily sought to escape the dilemma in a world view involving the principles of love and strife. Love attracts and unites, strife repels and divides, he said, and the everlasting elements of earth, water, air, and fire are ruled by the sway of love and strife. "Nay, there are these things alone, and running through one another they become now this and now that, and yet stay ever as they are." From love and strife flow "forth the myriad species of mortal things, patterned in every sort of form, a wonder to behold."

*   *   *

Protagoras, a Thracian born in Abdera, was prominent among the Sophists who made their living by teaching the art of rhetoric. Skill in rhetoric and the talent to prove that black is white were (and still are) invaluable in the legal and political professions. Sophistry has other uses, as Samuel Butler the satirist said of those clergymen who switched their religious allegiance back and forth as Charles I, Cromwell, and Charles II moved in and out of office:

> What makes all doctrines plain and clear?
> – About two hundred pounds a year.
> And that which was prov'd true before
> Prove false again? Two hundred more.

Two hundred pounds in the seventeenth century was the annual income of a clerical living. Protagoras would have approved this game of musical chairs. He thought the gods probably did not exist, but prudence dictated that one hedged one's bets by worshipping at least one of them. He is famed for saying, "Man is the measure of all things," which contains considerable truth when used with care. Rejection of all absolute values led the Sophists into a philosophy of relativism. Ethical values, they argued, are purely relative; what is good in one society may be bad in another, what is right and proper for one person may be wrong and improper for another, and nothing is either right or wrong but thinking makes it so. Arguing thus, with their theory of social and ethical relativism, they fostered a hedonistic belief that pleasure and the gratification of personal desire are all that truly matters.

Socrates, a renowned philosopher who lived in Athens in the fifth century B.C., devoted his life to countering Sophist doctrine. The truth lies within us, he taught. Self-knowledge is wisdom, the doorway to serenity of mind, and we learn by searching within ourselves what is right and wrong. The Socratic method of inquiry consisted of asking questions, and step by step the interrogated person uncovers knowledge previously possessed unconsciously.

Scientists invent postulates and when the ensuing deductions are in accord with observations that is their sufficient reason until better postulates are found. The Ionians did not perplex themselves with the problem of why the postulated elements exist; instead, they asked how the elements work, evolve, and account for what is observed. Socrates explained to Cebes in the *Phaedo* why he felt that this failed to get to the heart of the problem: truth must be uncovered and cannot be invented.

In the Platonic universe the Mind, or cosmic demiurge, operated according to a plan that was fully known to the soul and was knowable by inward inquiry. Plato, at the Academy in Athens, had faith in the existence of a rational reality beyond the shadowy world of physical forms. Experiences are appearances, ideas are realities. To this day we are the bewildered heirs of this topsy-turvy doctrine.

*   *   *

Three great cosmic systems – Epicureanism, Aristotelianism, and Stoicism – dominated the Hellenic world and survive to this day imprinted in the cultures of modern societies. Each system combined philosophical, scientific, ethical, and religious elements.

The Epicurean system of endless worlds emerged in the fifth century B.C. from atomist ideas and later gained wide support in the Hellenic cities and city states. Epicureanism stressed the importance of the life sciences, rejected the gods as explicative agents in the natural world, and accepted the intellectual equality of men and women. It flourished for seven centuries and was then suppressed because of its atheistic rejection of the gods. It has reemerged in recent centuries and now forms the basis of the modern physical universe.

The Aristotelian system of etheric celestial spheres originated in Athens at the time of Aristotle, and centuries later was adopted by the Judaic–Christian–Islamic religions. Outfitted with theistic additions of Babylonian and Zoroastrian origin, updated with Stoic improvements, it evolved into the medieval universe that endured until the sixteenth century.

The Stoic system originated in Athens in the third century B.C. and stressed the significance of the natural sciences and the paramount importance of ethical principles. The Stoic universe consisted of a finite cosmos of stars surrounded by an endless extramundane void. The Stoic island universe endured in various forms until the early twentieth century, and its cosmology and its emphasis on science formed the basis of the Victorian universe.

*   *   *

The Epicurean universe began in the fifth century B.C. with the atomist ideas conceived by Leucippus, of whom little is known, and by his follower Democritus, who taught at Abdera in Thrace. From their thoughts and those of other philosopher–scientists emerged the concept of a universe of countless worlds distributed throughout infinite space. All worlds are composed of atoms, said the atomists, and the atoms differ in shape and size and consist of the same primary substance. The sensations of color, sound, smell, touch, and taste exist not in things themselves but in our sense organs. "By convention there is color, by convention sweetness, by convention bitterness, but in reality there are atoms and the void," said Democritus. All else is opinion and illusion. If the soul exists, it also consists of atoms. Most Athenian philosophers, including Socrates, Plato, and Aristotle, rejected the atomist theory and we are indebted to the Epicureans for preserving and developing atomist ideas.

Epicurus of Samos settled in Athens in the fourth century B.C. and founded the Epicurean school of philosophy – the first school to admit women students. The Epicureans (followers of Epicurus) adopted the atomist theory of numberless worlds strewn throughout an infinite universe. Each world, they said, consists only of atoms. Endlessly and freely the atoms move through the void, repeatedly colliding, occasionally aggregating, forming worlds that evolve and ultimately dissolve back into the atomic ferment. On each world life originates as primitive organisms and evolves to an advanced civilized state. To this day the sweep of the Epicurean vision grips the imagination.

In the epic poem *The Nature of the Universe*, in praise of Epicureanism, the Roman poet Lucretius wrote in the first century B.C.:

> But multitudinous atoms, swept along in their multitudinous courses through infinite time, by mutual clashes and their own weight have come together in every possible way and realized everything that could be formed by their combinations. So it comes about that a voyage of immense duration, in which they have experienced every possible variety of movement and conjunction, has at length brought together those whose sudden encounter normally forms the starting-point of substantial fabrics – earth and sea and sky and the races of living creatures.

Echoing Epicurus, Lucretius declared, "Bear this well in mind and you will immediately perceive that nature is free and uncontrolled by proud masters and runs the universe by herself without the aid of gods."

Epicureans thought that human beings are superior animals, and believed that the divine spirit existed not in the gods but in ourselves. They believed that the real pleasures in life stems from moderate living. Epicureanism flourished in the Greco-Roman world and finally perished with the spread of Christianity. A surviving manuscript of the Lucretian poem was found in 1417 hidden in an Eastern European monastery. It became widely known with the invention of printing using movable type in 1436 and contributed to the fall of the medieval universe.

*   *   *

The Aristotelian universe began with the two-sphere system popular among Athenian astronomer–philosophers in the early fourth century B.C. It consisted of little more than the Earth surrounded by a sphere whose inner surface was studded with stars. The Earth (the inner sphere) stayed motionless at the center of the universe, and the heavens (the outer sphere) rotated daily. Overhead, beneath the stars, the planets moved in their individual ways. At the Academy, under the leadership of Plato, intermediate spheres were added, and these new spheres, which supported the planets, rotated at various rates about inclined axes. To explain the motions of the planets the academicians transformed the two-sphere model into a many-sphere model of the heavens.

Aristotle of Stagira in northern Greece studied at the Academy. He became tutor at the Macedonian court to Alexander, a youthful firebrand who later became king of Macedonia. While Alexander was off conquering the Middle East, the Persian Empire, and lands farther east, Aristotle returned to Athens and founded his own school, known as the Lyceum. His lectures ranged widely, covering natural history, biology, physics, logic, politics, and ethics.

Aristotle took the many-sphere model and invested it with physical reality. The planets, including the Sun and Moon, in order of their geocentric distance, were the Moon, Mercury, Venus, Sun, Mars, Jupiter, and Saturn, and each had its supporting system of linked crystalline spheres. Altogether, fifty-four spheres were needed to make it work. It was a geometric, geocentric, finite universe extending to the outermost sphere of stars. The tireless rotations of the many spheres, said Aristotle, have persisted through eternity. It was of finite extent in space and of infinite duration in time.

In the Aristotelian universe the physical elements of earth, water, air, and fire were the constituents of the Earth and the sublunar region. The heavenly bodies and their supporting spheres consisted of a fifth element called ether. The natural motion of the physical elements was upward and downward, as they sought to find their proper place according to weight. The natural motion of the etheric element was endless rotation around the Earth. It is fitting, said Aristotle, that the physical elements of perishable forms should have imperfect incomplete motion, away from and toward the center of the universe, and this explains why the physical Earth does not rotate. Also, it is fitting that the etheric element of imperishable forms should have perfect circular motion, and this explains why the heavens forever rotate around the center of the universe. Generation and decay occurred only in the physical realm of the Earth and the sublunar region. In the etheric realm, above the sublunar region, everything remained changeless and perfect.

Comets and whatever marred the perfection of the heavens were no more than atmospheric phenomena. This belief persisted for two thousand years, and whenever a new star flared brightly in the sky, observers shook their heads in disbelief.

Claudius Ptolemy, an astronomer and mathematician at the Museum of Alexandria in the second century A.D., did for astronomy what Euclid (at the Museum four centuries earlier) had done for geometry. He brought together the astronomical observations made in previous centuries, and in his *Almagest* (meaning "Great System" in Arabic) he used epicyclic geometry to explain the motions of planets. The result was a geometric marvel that endured for fourteen hundred years until replaced by the revolutionary works of Copernicus, Kepler, and Galileo.

The final form of the Aristotelian universe, as presented by Ptolemy, failed to incorporate many developments in Greek science. It rejected the notion of atoms, the suggestion by Democritus that the Milky Way is an agglomeration of stars, the proposal by Heracleides that the Earth rotates daily, and the theory by Aristarchus (accepted by Archimedes) that the Earth rotates daily and revolves annually about the Sun. Aristarchus of Samos in the third century B.C., inspired by

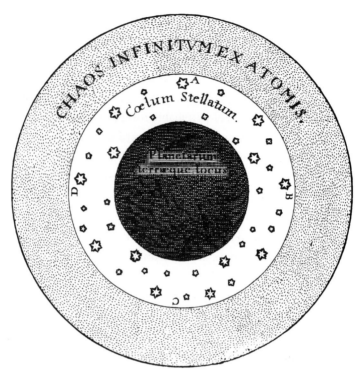

The "world system of the ancients," according to Edward Sherburne (1675). This illustration, representing the Stoic system, shows an inner sphere of planets, surround by a sphere of stars, which in turn is surrounded by infinite empty space.

a Pythagorean idea, showed how the apparent seesaw motion of the planets could be explained. If we assume that the planets, including the Earth, revolve around the Sun, he said, then all other planets as seen from Earth will exhibit the observed forward and backward motion. But the idea of a Sun-centered system was not generally accepted. Its revival in the sixteenth century formed the basis of the Copernican revolution.

*   *   *

Zeno of Citium, born in the fourth century B.C., founded in Athens the Stoic school of philosophy. He lectured in a roofed colonade (called a *stoa*) to all who cared to listen, and his philosophy and ethics, later elaborated and known as Stoicism, appealed to all classes, from slaves to aristocrats. He exalted duty, justice, and self-reliance, and condemned disloyalty and injustice. We may imagine this strange man calling out to all who passed by: "Stand by those you cherish and love. Be brave in the face of adversity. Weep not for thou art strong! Gaze on it all and be not amazed or afraid for the soul has seen it many times before." Stoicism spread throughout the Roman Empire and its ethical ideals of duty, honesty, and justice, expressed in highest form in the writings of Seneca and Marcus Aurelius, now permeate Western cultures.

Stoic metaphysics taught that proper understanding requires the study of the whole rather than just its bits and pieces. The Stoics firmly believed in fate. All that happens and will happen is predestined. They believed that the Mind, manifesting through gods and mortals, governs the universe. Some said the stars were alive and the universe was a living organic whole. The Stoic cosmos of a cluster of stars surrounded by an infinite void ended finally at the beginning of the twentieth century.

*   *   *

The intellectual giants of the Hellenic world created cosmic systems that have since shaped the outlook of almost all human beings. They turned the tide against the mythic universe and reactivated the world in ways that puzzle us to this day. They restored the spirits of the age of magic that now masquerade as electrons, protons, neutrons, quarks, gluons, gravity, electricity, magnetism, wavefunctions, potentials, inertia, momenta, energies, pressures, and the rest, disciplined by a Pythagorean numerical harmony.

We feel tempted to think that all Greeks reasoned from conjectured principles, like the Socratics, and were theorists, not experimenters and critical observers. But this temptation must be resisted. From Thales investigating the properties of water, Hecataeus formulating practical rules in geography, Pythagoras studying the resonances of vibrating strings, Hippocrates discovering the methodology of medicine, Aristotle dissecting the biology of life, Archimedes inventing levers, mechanical contrivances, and a method of measuring density, Eratosthenes measuring the diameter of the Earth, to the pumps, steam engines, and research projects at the Museum in Alexandria, a history of empirical inquiry unfolds without which there could have been no science.

*   *   *

What might have happened to the human race if there had been no science? Let us imagine what might have happened to the human race in the last two thousand years if there had been no science. Suppose that Thales had not lived. Possibly only a small fraction of all people now inhabiting the globe would be alive. Of these, most would live as serfs or slaves in a mythic universe, governed by god-inspired despots, with death by disease and malnutrition the common lot. If this seems an exaggeration, throw in all the Ionian philosopher–scientists and include the Pythagoreans. Little doubt can remain that life would be vastly different and much less pleasant than it is for most people now living in the modern physical universe. Of all the miracles of the mythic universe, the most remarkable was the emergence of science.

5 The Medieval Universe

"I drew these tides of men into my hands and wrote my will across the sky in stars," wrote Lawrence of Arabia in *The Seven Pillars of Wisdom*. Human tides have washed across the globe, crushing nations and carving out empires, led by god-possessed men who sought to write their will across the sky in stars. One such leader was Alexander the Great, who crossed the Hellespont with his cohorts in the fourth century B.C., subjugated Asia Minor and Egypt, vanquished the armies of the Persian Empire, quelled the turbulent forces of Afghanistan, crossed the Hindu Kush, and invaded and defeated the nations of the Punjab.

Eastward flowed Hellenic philosophy and science in the wake of Alexander's conquests; westward flowed oriental philosophy and religion. Westward into the Mediterranean world came the glorious Ahura Mazda – the Zoroastrian Lord of Light embattled with the Lord of Darkness – bringing the belief that the soul is divine and the worship of gods other than the true god a sin. Westward into the Roman legions came the religion of the dying and resurrected martyred god, the triumphant Mithras, bringing the sacramental eating of the flesh of the god and the notions of forgiveness and redemption. Westward came the Babylonian stories of the creation and the flood, the Persian stories of heaven and hell, the last day of judgment, and the resurrection of the dead, all of which shaped the theology and philosophy of the Greco-Roman world in preparation for the rise of Christianity and Islam.

Centuries later in the hinterland beyond the Volga a nation of Huns erupted in pandemonium, attacked by fearsome nomadic Avars. Hordes of dislodged Huns swept through the empires of the Ostrogoths and Visigoths. The Goths fled before the storms of arrows and crossed the Danube into the Roman Empire. The fleeing Goths pressed on the Vandals, also a Germanic people, who joined in the pell-mell rush to escape the tide of terror. After the death of Attila the Hun (the "Scourge of God"), the Huns were defeated and dispersed by combined Gothic, Celtic, and Roman armies. The crazed Vandals, who had lost their homes, wives, and children, sacked Rome, then fled again before the Goths and established a kingdom in North Africa from where they harried the Mediterranean with pirate fleets until suppressed by Byzantine forces.

Theodoric the Goth became king of Italy toward the end of the fifth century and sought to restore order amid the ruins of the Roman Empire. According to Edward Gibbon, in his history *The Decline and Fall of the Roman Empire*, the defeat of the empire was the "triumph of Barbarism and Religion." Historians now offer other views: political corruption, military anarchy, economic chaos, bureaucratic oppression, excessive taxation, and breakdown in the judicial system had destroyed the empire from within long before the barbarians gained their victories.

In the seventh century, the Arabs poured out of their deserts and founded the Islamic Empire that stretched from Spain to India. Islam (meaning "piety") proclaimed the power and glory of the "One God." Trade thrived by land and sea and linked together an empire of unusual religious tolerance. Judaic scriptures formed the historical foundation of the new religion. Nestorian teachings that Christ was an inspired prophet but otherwise an ordinary human being and his mother an ordinary mortal influenced the formulation of Islamic doctrine, and to this day the prophet Mohammed is looked upon not as God but as the inspired vehicle of the voice of God.

For thousands of years nomadic Mongolian and Turkic people had periodically sallied forth from the steppes of Central Asia. To withstand their benighted assaults, the Chinese in the third century B.C. built the Great Wall on their northern frontier. Once more, under the leadership of Genghis Khan in the thirteenth century, the descendants of the "blue wolf and gray dove" rallied, and again warrior horsemen swept southward and westward. The earth trembled to the thunder of hoof beats and the sky darkened with the sack of cities. The Mongolian Empire of Kublai Khan, grandson of Genghis, covered more than a quarter of the land surface of the globe. Along the Silk Road traveled intrepid European adventurers, including the young Marco Polo, who were dazzled by the unsuspected magnificence of oriental civilization.

With the death of Kublai Khan, at a time when Europe stood at the brink of being engulfed, the empire broke into a conflict of warring armies. "A monstrous and inhuman race of men has appeared from the East," cried an Arab ambassador seeking help from the West. But to no avail. The Byzantine Empire of Egypt, Asia Minor, and Balkan Peninsula was swept into the Turkic Ottoman Empire, and Constantinople was finally defeated in the fifteenth century. The great Byzantine bastion that defended Europe for more than a thousand years had fallen. Out of the devastation, along caravan routes and in the wake of armies, came disease-infested rats. More than half the populations of Asia, North Africa, and Europe died in the plagues that followed.

In scant words this is the historical background to the medieval universe of the Middle Ages.

*   *   *

The medieval universe – the Eternal City and dream of Saint Augustine – reached its zenith in the high Middle Ages of the twelfth to fourteenth centuries and was the last and grandest of the mythic universes.

The religious rudiments of the medieval universe were the Hebraic scriptures and gospels. The history of the world had unfolded according to a divine plan whose major events were the Creation, Fall, Flood, Election of the Israelites, Exodus from Egypt, works of the prophets, Exile in Babylon, Incarnation, Crucifixion, Resurrection, and the Day of Judgment. All other events, such as the Egyptian and Mesopotamian civilizations and the Roman Empire, served as accessories in the implementation of the plan.

Man and woman in the beginning were made perfect, but because of their original sin of willful disobedience they fell from grace into a state of spiritual deprivation. God sent his only son, the Redeemer, to show the way of atonement and salvation. The wrath of God could be averted by sacrifice and appeal to mercy, but the original sin must stay forever unforgiven until the last day of judgment when all persons will receive their just deserts: the wicked condemned to everlasting torment, the good restored at last to the spiritual grace of the first man and woman.

It was inconsistent with doctrine that a Christian should live in slavery, possessed body and soul by a human master rather than by God, and with the spread of Christianity into Europe in the early Middle Ages (the third to the eighth centuries), slavery retreated and almost vanished.

Benedictine monks in the sixth century launched a large missionary enterprise that established monasteries and schools in Western Europe. The Benedictines taught not only the elements of orthodox doctrine, but also the trivium consisting of grammar, logic, and rhetoric that in earlier centuries formed the basis of the curriculum in Roman schools. After Charlemagne (the eighth century), the monastic schools taught also the quadrivium consisting of arithmetic, astronomy, geometry, and music. The trivium and quadrivium together comprised the liberal arts. Roman compilations, such as Pliny's *Natural History*, served as supplementary texts. The works of Boethius, a renowned scholar in the early Middle Ages, whom Theodoric executed on a charge of conspiracy, formed part of the curriculum. While awaiting execution, Boethius wrote the *Consolation of Philosophy*, and this work and his translated fragments of Euclid, Aristotle, and Ptolemy contributed to the intellectual recovery of Europe.

Meanwhile, under the rule of the caliphs, the arts, crafts, and sciences thrived. Greek, Jewish, Persian, and Indian scholars flocked to centers of learning in Baghdad, Damascus, Cairo, and Cordoba where libraries were stacked with ancient manuscripts. Europe slowly awoke, bestirred by the impact of new lifestyles and novel thoughts.

The stirrup transformed European feudalism and made possible an aristocracy of mounted warriors in an age of chivalry. Legends tell of damsels in distress but not of the skilled artisans who manufactured and maintained the knightly armor. The introduction of the heavy wheeled plough, padded horse-collar, nailed horseshoe, and storage of hay revolutionized agriculture and greatly increased the production of food. The standard of living rose, populations grew, and the barbarian vernaculars of Latin evolved into the romance languages French, Italian, Spanish, Portuguese, and Rumanian.

With slavery banished, Europe began to throw off the traits of ancient living. Mind-dulling harsh toil slowly disappeared in a society sustained by the skills of artisans and the investments of financiers. Cistercian monks, living in mechanized communities and using labor-saving methods, pioneered the technology revolution. Rivers, winds, and tides supplied power to water wheels, windmills, and tidal mills. Mechanisms – some copied from the Chinese – consisting of transmission shafts, driving belts, gear trains, flywheels, cranks, cams, springs, and treadles became widespread. In a mechanical-crazy Western Europe of the high Middle Ages, mills busily ground, mixed, crushed, sawed, pulped, and operated bellows and trip hammers for forging iron. It became an age that also built the great cathedrals.

Mechanical clocks – the product of advanced technology – appeared in the late thirteenth century. In *Medieval Technology and Social Change*, Lynn White writes,

> Something of the civic pride which earlier had expended itself in cathedral-building was now diverted to the construction of astronomical clocks of astounding intricacy and elaboration. No European community felt able to hold up its head unless in its midst the planets wheeled in cycles and epicycles, while angels trumpeted, cocks crew, and apostles, kings, and prophets marched and countermarched at the booming of the hours.

And we should not forget the invention of spectacles, which in this age of technical genius extended the working life of scholars, artists, and craftsmen.

A revolution had occurred unlike any in history. The skills of artisans were no longer the monopoly of the courts, but were used for the benefit of many in a mechanized society. The Middle Ages, long referred to as the Dark Ages by historians trained in the liberal arts who had low regard for the "servile" arts, were a time of social change of immense importance. Ordinary people, skillful and industrious, discovered they had social value.

* * *

Already by the end of the ninth century Western Europeans knew the Earth was a sphere and that the universe, contrary to the Hebraic scriptures, had geocentric symmetry. Inspired by Islam with its foothold in Spain and Sicily, inquisitive monks thirsting after new knowledge began to take an interest in the legacy of classical antiquity. Arabic manuscripts, when translated into Latin, created intellectual unrest, and tantalizing fragments of Euclid and Aristotle triggered trains of novel thought.

An age of translations began. Words of Arabic origin, such as algebra, alkali, azure, camphor, cipher, borax, elixir, jasmine, jute, saffron, sherbet, zenith, and zero, gained currency. Schools of scribes in the twelfth and thirteenth centuries busily translated into Latin whatever Greek manuscripts they could find. The flood of new knowledge overflowed the monastery and cathedral schools, and much of it stayed in the hands of the translators, who became professional educators. Communities of learned teachers at Bologna, Padua, and Salerno taught the liberal arts, medicine, and the law. These communities were the early universities, to which students traveled from far and wide. Knowledge and learning became the surest route to social promotion and high office. Students paid their fees to the professors and formed unions to ensure they got their money's worth; the professors in turn formed academic unions, or faculties, which regulated the award of bachelor degrees (licenses to practice) and doctorates (licenses to teach).

The universities of France and England developed a formal structure, and functioned under royal charter and papal authority. Students were subject to canon law and exempt from common law. Colleges (endowed halls of residence) promulgated rules of decorous behavior and were a conspicuous feature at the universities at Oxford and Cambridge. Control at the University of Paris was vested in the chancellor, a dignitary of the Church, and the faculties of theology, medicine, law, and arts each had a presiding dean (a Church dignitary). The faculty of arts, the largest, taught the trivium and quadrivium, both greatly enlarged by the influx of new knowledge. The curriculum at Paris in the mid thirteenth century included courses on astrology, weather, physics, animals, plants, ethics, sense and sensibles, sleep and waking, memory and remembering, and life, death, and the soul. Students worked hard; a master's degree in arts took usually six years of study, followed by eight more years for a doctorate in theology. Of the approximately seventy universities scattered around Europe in the late Middle Ages, almost all followed the Paris model with theology as the leading subject. Charles Haskins in *The Rise of the Universities* remarks, "We are the heirs and successors not of Athens and Alexandria, but of Paris and Bologna."

Hitherto, elements of Roman law had remained entangled in Gothic codes. With the revival of classic learning came the study and practice of Roman law and the restoration of judicial torture as a means of determining guilt and innocence. In the witch-craze of the Renaissance, hundreds of thousands of victims were tortured in accordance with the principles of Roman judicial inquiry. Gothic and canon law were preferred to Roman law in England and, particularly after Magna Carta, judicial torture was used only for acts of treason, consistent with Gothic tradition. Witches were burned only during the reigns of Roman Catholic kings and queens under the direction of papal Roman law.

At first, the universities were dominated by the mendicant orders – Franciscans and Dominicans – whose members ranked among the most learned thinkers of the Middle Ages. The Franciscan monk Roger Bacon typified the fluidity of thought of this period. Wholeheartedly he embraced Aristotelian empirical science and sought to unravel the secrets of nature. He foresaw the outcome of the technology revolution and predicted ships that would move without sails or rowers, vessels capable of exploring the bottom of the seas, flying machines, and prophesied, "wagons may be built which will move with incredible speed and without the aid of beasts."

*   *   *

The most influential of the new ideas in the universities came from the works of Aristotle; they created intellectual excitement, opening the door to a world of rational inquiry. Averroes, an Arab scholar of Cordoba in Spain, showed how Aristotelian knowledge could be harmonized with Islamic beliefs. Moses Maimonides, a learned rabbi also of Cordoba, did much the same for Judaic beliefs. In the thirteenth century, Thomas Aquinas, a black-robed Dominican, followed in their footsteps and demonstrated how Christianity could be accommodated within a modified Aristotelian system. Aquinas and other learned divines took the greatest of all contemporary themes – the narrative of sin and salvation – and wove it into the fabric of Aristotelian cosmology. From their work emerged the medieval universe in final form.

"In the beginning," according to Genesis, "God created the heavens and the earth." The medieval universe, unlike the Aristotelian system, had a beginning, and was created by God to serve a specific end. Beyond the sphere of fixed stars lay the *primum mobile*, a primary sphere introduced by the Arabs, that divine will maintained in constant motion; and beyond the *primum mobile* extended the empyrean, a realm of purest fire, conceived by Saint Anselm, where God dwelt. The ascending planetary spheres accommodated a hierarchy of angelic beings whose degree of divinity increased

God creates and maintains the universe. From Martin Luther's *Biblia*, published by Hans Lufft, Wittenberg, 1534.

with altitude. Aerial and daemonic beings trod a less-orderly measure in the sublunar sphere. Earth was the home of mortal life and its earthly elements formed the perishable vessel of the immortal soul.

Hell, located in the bowels of the Earth, was where the wicked went to be eternally punished. Above the Earth's surface and beneath the sphere of the Moon lay purgatory where spirits were purged before ascending farther. Guarded by Angels, the lunar sphere served as the entrance to the higher spheres of heaven. Beyond the celestial spheres, above the *primum mobile*, in the empyrean, God looked down and watched over his creation. By compromise the learned fathers combined reason with faith and gained a universe of monumental elegance.

Aquinas in an Age of Faith used reason to justify faith. Voltaire in an Age of Reason half a millennium later used faith to justify reason. Yet the difference between Aquinas and Voltaire is less than we might think. Carl Becker in *The Heavenly City* writes, "What they had in common was the profound conviction that their beliefs could be reasonably demonstrated." Both believed they lived in a universe of rational meaning. We nowadays live in a universe where the question of its meaning is without meaning. Reason in faith has gone and faith in reason is itself without reason,

Every aspect of the medieval universe had meaning. Human beings occupied the most prominent of all places: the Earthly stage, with the spotlight beamed on them as the leading actors in a drama of cosmic proportions. Blessed by religion, rationalized by philosophy, and verified by geocentric science, the medieval universe gave meaning and purpose to life on Earth. Most persons living in that age could grasp the essentials of their universe and felt impelled to worship its creator. "Other ages have not had a Model so universally accepted as theirs, so imaginable, so satisfying to the imagination," wrote C. S. Lewis in *The Discarded Image*.

* * *

Many of us live in cities or towns where the night sky is lost in a glare cast by electric lights. Even when the night sky is seen clearly, we glance at it casually, for it means little to us. When we think about it, we know that we look out to vast distances in a universe that is dark and mostly void. This was not so in the medieval universe. People looked up unhindered by the glare of electric light to a celestial panorama of immediate significance, resonant with the choirs of heaven. They saw a universe radiant with the bright blue light of heaven. The "bright blue firmament" in the Middle Ages was a fact, and the blueness of the daytime sky was not scattered sunlight by the atmosphere but the light of heaven. The higher the celestial sphere, the more dazzling became the ethereal light. At nighttime, according to medieval scholars, the blue light of heaven could not penetrate the Earth's shadow. Demons from nether regions arose and roamed freely in the darkness. The alternation of day and night testified to the unending struggle between the powers of light and darkness, good and evil.

To people of those times the magisterial medieval universe seemed immense in size. Lewis writes in *The Discarded Image*, "For thought and imagination, ten million miles and a thousand million are much the same." In the modern physical universe the Earth seems very small, but so does everything else, even the galaxies. The medieval universe with its outer boundary at finite distance made the Earth's smallness vividly apparent. "To look up at the towering medieval universe," said Lewis, "is much more like looking at a great building. The 'space' in modern astronomy may arouse terror, or bewilderment, or vague reverie; the space of the old presents us with an object in which the mind can rest, overwhelming in greatness, but satisfying in its harmony." Universes always amaze their inhabitants by their vastness. Amazement today at the extent of the physical universe echoes the amazement in the Middle Ages at the extent of the medieval universe.

The fantasy of journeying away from the Earth as a space traveler, ascending through the celestial spheres, and then looking back and seeing the Earth as a distant and tiny orb, originated in the first century B.C. in the works of Cicero. The fantasy was often used in the Middle Ages and served to emphasize the grandeur of the heavens and the relative smallness of the Earth. Dante employed it with great effect.

*　　*　　*

The Neoplatonic idea of God at "the center of the world," elaborated in the mystical writings of Pseudo-Dionysius (an unknown disciple of Proclus in the fifth century), never entered the mainstream of Christian doctrine. Unlike the classic geocentric picture, the theocentric Pseudo-Dionysian universe had inverted structure. God occupied the center of the universe, as seemed fitting to Gnostics and Neoplatonists, and was surrounded by angelic spheres. Beyond the outermost sphere lay darkness where human beings dwelt.

Even the most pious cleric found it difficult to ignore the astronomical fact that the Earth and not God had central location, and the theocentric universe failed to gain wide acceptance. Christianity and Islam were both nurtured on the Platonic concept of God as omniscient and omnipresent, and neither religion could accept the idea of God confined to a fixed point. In the throne verse of the Koran we read, "His throne is as wide as heaven and earth, and the preservation of them wearies Him not, and He is the Exalted, the Immense."

The ingenious Dante Alighieri in the early fourteenth century, with artistic license, succeeded in bringing together within a unified universe the geocentric and theocentric systems. In the *Divine Comedy* ("divine" was added later, and "comedy" means a happy ending), Dante placed the angelic spheres within the empyrean in such a way that the celestial and angelic spheres mirrored each other.

It is easy to construct a simple model that illustrates Dante's universe. Take a large disk of white cardboard (size of a dinner plate), and on one side mark in the center a point to indicate the Earth. Draw around this point eight concentric circles of increasing size to represent the celestial spheres (the Moon, Mercury, Venus, Sun,

The *Empyrean* by Gustave Doré (1832–1883), showing Dante and Beatrice gazing upon the theocentric world of angelic spheres from the rim of the antipodal geocentric world of celestial spheres.

Mars, Jupiter, Saturn, and Stars), and let the rim of the disk be the *primum mobile*. On the other side mark in the center a point to indicate God. Around this center draw again eight concentric circles of increasing size to represent the angelic spheres (the Seraphim, Cherubim, Thrones, Dominions, Virtues, Powers, Principalities, and

Archangels), and let the rim of the plate in this case be the sphere of the Angels. On one side of the plate we see the geocentric world of celestial spheres; on the other side the theocentric world of angelic spheres, and mediating between the two at the rim are the Angels occupying the *primum mobile*. This model, in which Earth and God are the antipodes of a symmetric universe, shows how Dante brought into harmony the material and spiritual realms.

In his imaginary journey, as recounted in Canto 28 of *Paradise*, Dante leaves the Earth and ascends through the celestial spheres to the rim of the universe and sees on the other side a panoramic view of heaven:

> One point I saw, so radiant bright,
> So searing to the eyes it strikes upon,
> They needs must close before the searing light.
>
> About this point a fiery circle whirled,
> With such rapidity it had outraced
> The swiftest sphere revolving round the world.
>
> This by another was embraced,
> This by a third, which yet a fourth enclosed;
> Round this a fifth, round this a sixth I traced.

... and so on. While standing at the rim he sees before him a brighter world similar in arrangement to the one left behind.

Dante's remarkable synthesis made very little impact on theology and cosmology. The notion that God could be geometrized, while permissible in flights of poetic fancy, was otherwise impermissible, for it imposed geometric limitations on the form of God. The standard model, to which Dante subscribed in his other works, consisted of a set of angelic spheres superposed on a corresponding set of celestial spheres. Angels of different kinds populated the heavens and provided, according to some accounts, the motive force that maintained the rotation of the celestial spheres. The Angels occupied the sphere of the Moon, the Archangels occupied the sphere of Mercury, and so forth, to the Seraphim who occupied the *primum mobile* and were closest to God.

*   *   *

Most astronomers from the Babylonians to the Elizabethans regarded themselves as astrologers. It should be understood that "astrology" had not its present meaning. It was, as the name implies, the science of planets and stars, their eclipses, emanations, and influences on one another. Geoffrey Chaucer's *Treatise of the Astrolabe*, dealing with celestial observations, was for a long time referred to as an astrological work. In the late Middle Ages and until the sixteenth century, astrology meant literally the science of celestial phenomena; whereas astronomy was the art of naming and identifying of stars and constellations.

After Alexander the Great had opened the floodgates to oriental cults and mystery religions, astrology became linked with the temple cults of astrolatry (worship of astral bodies) and with the arcane arts of astromancy (astral divination and horoscopy). In recent centuries the science of astrology has been renamed "astronomy," and astromancy under the name of "astrology" has concerned itself with the effect of planetary movements on the affairs of human life.

In the first half of the sixteenth century the Swiss physician Paracelsus and the Flemish anatomist Vesalius discarded the mythically encrusted medical lores of Galen and Avicenna and laid the foundations of modern medicine; Copernicus at about the same time discarded the geocentric Ptolemaic system and started a new age in astronomy; and chemistry, divorced from alchemical and medical lore, began as a natural science with Robert Boyle's *Skeptical Chemist* in the mid-seventeenth century.

*   *   *

The medieval universe from yet another viewpoint was the Great Chain of Being. The Neoplatonists developed the notion that the world of living creatures consisted of countless graduated forms of life. This view of the living world was popular in the late Middle Ages and greatly influenced European thought until the nineteenth century. Link by link, the great chain of sequential lifeforms descended from human beings through beasts and plants to insensible matter, and link by link ascended through angelic forms to the throne of God. Human beings were the central link connecting the brute and angelic realms. All known and imaginary species fitted into the grand arrangement and no gaps could exist to blemish its perfect continuity.

Arthur Lovejoy explains in *The Great Chain of Being* how this theme captured the imagination of Europeans and made an indelible impression on their literature and art. In *Essay on Man*, Alexander Pope wrote:

> Vast chain of being! which from God began,
> Nature aetherial, human, angel, man.
> Beast, bird, fish, insect, what no eye can see,
> No glass can reach; from Infinite to thee,
> From thee to nothing. – On supreme powers
> Were we to press, inferior might on ours;
> Or in the full creation leave a void,
> Where, one step broken, the great scale's destroyed;
> From Nature's chain whatever link you strike,
> Tenth, or ten thousandth, breaks the chain alike.

Lyrics voiced scientific beliefs. The connecting links forged by the Creator disallowed any possibility of evolutionary change. Were only one species to change or disappear, the severed chain would crash to the ground. This grand picture of a biological chain of immutable life-forms decreed by God was what evolutionists had to struggle against in the nineteenth century. The chain moored the living world to God and secured for human beings a central position of cosmic importance.

The Middle Ages also made articulate the principle of plenitude implicit in Judaic and Christian doctrine. The notion of plenitude sprang from the belief that a benevolent Creator had given to human beings for profit and exploitation a richly endowed Earth. This sentiment is expressed in the Eighth Psalm: "Thou has made him a little lower than the angels, and hast crowned him with glory and honour. Thou madest him to have dominion over the works of thy hands; thou hast put all things under his feet." Sheep, oxen, beasts of the field, fowl of the air, and fish in the sea existed solely for the benefit and pleasure of human beings.

The principle of plenitude dovetailed neatly into the Great Chain of Being. The Earth possessed unlimited wealth of every possible kind, and displayed in profusion all possible forms of life with no gaps or missing links in the great chain. Land, sea, and air necessarily teemed with life in inexhaustible supply, and depletion of any species to the point of extinction was inconceivable, for "missing links" (a pre-Darwinian expression) would imply a makeshift creation. To doubt the existence of plenitude was equivalent to doubting God's munificence, and not consume to the utmost whenever possible was equivalent to rejecting God's gifts. Belief in plenitude and the right to consume to excess drove European nations to the uttermost limits of human effort; civilizations toppled before their fierce hunger, and their imperial empires of merchant adventurers girdled the globe.

The myth of plenitude, which lies at the roots of modern political ideology, was the motivating force, and whenever anything showed signs of extinction, hunters, trappers, whalers, fishermen, lumbermen, miners, farmers, explorers, fortune seekers, financiers, politicians, clergymen, and the consuming public were overtaken by surprise and unable to believe that exhaustion had actually occurred.

*   *   *

Already in the middle of the thirteenth century alarmed ecclesiastics were expressing concern that the conciliation of Christianity with Aristotelianism had gone much too far. The geocentric universe was all very fine, but if it placed constraints of any kind on the power of God, or if it meant that God existed only outside the *primum mobile*, or that God could not move the Earth, or could not create worlds other than the Earth, if he so willed, then it was false and contrary to orthodox doctrine.

Etienne Tempier, Bishop of Paris, issued in 1277 the famous 219 condemnations anathematizing "the execrable errors which certain members of the faculty of Arts who have the temerity to study and discuss in the schools." All discussions tending to limit the power of God were roundly condemned, for God was to be apprehended by faith and not intellectual conceit. God had unlimited power, said Tempier, and it was inadmissible to suppose that God was circumscribed by boundaries. The empyrean may indeed be the realm closest to the throne of God, as Anselm had proposed, but it must be realized that God dwelt everywhere and was not bounded in any way by the necessities of philosophy. The bishop's strictures put a stop to all speculations that might appear to diminish the concept of God, and one consequence of this dampening of Aristotelian enthusiasm was the coolness the *Divine Comedy* received when published a few decades later.

The condemnations of 1277 stand as a landmark in the history of cosmology. By asserting that God is neither here nor there but everywhere, they redirected inquiry and prepared the way for the Copernican revolution and the infinite universe. They made evident that the finite geocentric Aristotelian universe was too restrictive and could not accommodate an unbounded God of indefinite and perhaps infinite extent. In the years that followed, the universality of God became more and more reflected in the attributes of the created universe. If God indeed was boundless, why should not the universe – the handiwork of God – also be boundless? If God was without conspicuous location, why should human beings claim for themselves the privilege of central location? Theological concepts on the nature of God inspired secular concepts on the nature of the created universe.

*   *   *

The medieval universe has gone and we are left trying to pour its old wine into the new bottles of contemporary cosmology. In our social institutions, languages, and judgments of right and wrong, in our domestic life and everyday conversations, in our mechanical and labor-saving practices, in our numbers, manners, greetings, letters, nursery rhymes, and superstitions, in our ceremonies, rites of passage, courtships, courtesies, charities, beliefs in fair play, notions of honor and decency, and expressions of love can be found the enduring remnants of the medieval universe.

Though we feel lost in the modern and seemingly meaningless physical universe, deep down in our personal worlds we think as medieval people, and find comfort in the old beliefs. In the West we hold to the values and virtues of the medieval universe that lasted for a thousand years.

6 The Infinite Universe

Etienne Tempier, Bishop of Paris, roundly condemned all who dared to trifle with the power of the supreme being. Scholars and divines were free to admit reason into matters of faith provided full acknowledgment was made to God as an all-powerful being free of self-contradiction. Here was the Trojan Horse, introduced by the well-intentioned bishop, from which sallied forth in years to come thoughts that would topple the towers of the medieval universe.

*　　*　　*

Professors at Oxford and Paris in the fourteenth century made great progress in clarifying the nature of space, time, and motion. William Heytesbury and his colleagues at Merton College defined velocity and acceleration and then succeeded in calculating by graphical methods the distance traveled in an interval of time by a body having constant acceleration. William of Ockham participated in these studies while fighting a battle against needless abstractions. His celebrated principle of theoretical parsimony – known as Ockham's razor – states that in the use of concepts "it is foolish to accomplish with a greater number what can be done with fewer."

Jean Buridan, a professor at Paris and formerly Ockham's student, revived the notion of impetus that can be traced back to Hipparchus in the second century B.C. and is now referred to as momentum. According to Buridan, impetus is proportional to the velocity of a body and also its quantity of matter (now referred to as mass), and the impetus of a thrown body maintains the body in a state of motion. Aristotle, lacking the notion of impetus, supposed that rest is the natural state of all bodies, and a moving body must be pushed or pulled constantly by a force and will come to rest immediately when the force vanishes. Walking and swimming are examples of motion that must be maintained by continual effort. But the planets move freely, for they encounter no resistance, argued Buridan, and therefore need not be continually pushed or pulled along in their orbits by angelic forces. Instead, they move of their own accord because of the impetus imparted to them by God in the beginning.

Buridan promoted the idea that all bodies fall similarly, and when bodies of different weight are dropped side by side, they reach the ground simultaneously. This idea had been mooted by Philoponus of Alexandria in the sixth century, and after its revival by Buridan was adopted by other scholars, such as Leonardo da Vinci, and successfully tested by Simon Stevin of Belgium in the sixteenth century. Allowance must be made for the fact that air resistance introduces complications; feathers, for instance, fall as fast as stones in a vacuum but not in the resistive atmosphere.

From the fifteenth to the seventeenth centuries the universities added little more to the advance of science. They lagged behind, mired in the sticky problems of reconciling Aristotelian doctrine with religious dogma. Instead, it was the theologians, Bishop Oresme, Cardinal Cusanus, and Canon Copernicus, working outside the universities, who overturned the Ptolemaic system in what is called the Copernican Revolution.

* * *

Bishop Nicole Oresme of France in the fourteenth century had much to say about the unlimited power of God. "Motion is perceived by ordinary mortals," he said, "only when one body alters its position relative to another." Mortals perceive only relative motion, but God knows always the true absolute motion. We observe the heavens turning around the Earth, but only God can tell whether the Earth is still and the heavens revolve, or the Earth rotates and the heavens are still, and neither of the two is beyond the power of God. Our impression that the Earth is still might easily be wrong, for do not sailors in a moving ship have the impression that their ship is stationary and the sea and shore move? If the Earth moves, then everything on its surface, including the seas, the atmosphere, and ourselves, moves with it and shares in its impetus, and the old argument claiming that the Earth is necessarily at rest is false, and limits the power of God. Who knows whether one or many worlds were created? The inhabitants of other worlds, if such worlds exist, said Oresme, may also have the impression that they occupy the center of God's creation.

Oresme likened the universe to a delicately adjusted clock, and in his thoughts we see the first stirrings of a new world in which the geometric marvel of the ancients would be transformed into the clockwork marvel of the Newtonian world system.

The budding ideas of the bishop flowered in the fifteenth century in the fertile mind of Cardinal Nicholas of Cusa. In his work *Of Learned Ignorance*, the cardinal said, "every direction is relative," and "wherever a man stands he believes himself to be at the center." The ancients formed the opinion that the Earth is stationary and at the center of the universe. But has not God infinite power and wisdom, and apart from what mere mortals think, the Earth may move and not be at the center.

To suppose that the universe is finite with only one center, said the cardinal, when God is infinite and everywhere, is an inference unworthy of God's wisdom and creative power. More likely, "the absolute infinity of God has its counterpart in the infinity of the world as an image," he argued, and consequently "the universe has its center everywhere and its circumference nowhere."

Belief that God was unlimited and therefore infinite and everywhere led Nicholas of Cusa to the conclusion that the created universe was correspondingly infinite and its center everywhere. Wherever one stood in the universe, the stars would spread out much the same in all directions. God is unbounded and equally the same everywhere, he said, and the created universe similarly is unbounded and equally the same everywhere.

Nicholas of Cusa, theologian, scientist, and statesman, had an active and inventive mind. He advocated the counting of pulse beats

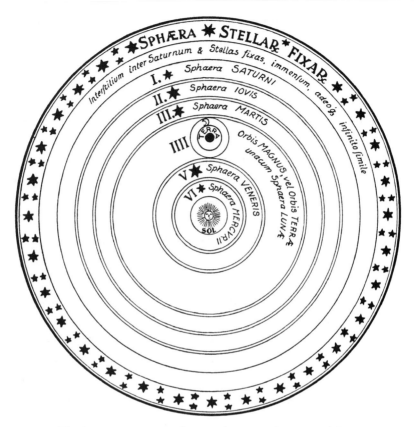

The Copernican system, showing the Sun at the center of the universe, encircled by the planetary orbits, and surrounded by an outermost sphere of fixed stars.

as a diagnostic aid in medicine, developed spectacles for the correction of nearsightedness, used jets of water to measure the passage of time, and claimed that plants draw on the atmosphere for nourishment. He hit also on the momentous cosmological argument that if the starry heavens are the same in every direction as seen from the Earth, and the universe has its center everywhere, then the starry heavens must appear much the same in every direction as seen from all other places. The logical conclusion was that all places in the universe are much the same; for if God is unbounded and equally the same everywhere, then plausibly the universe itself is unbounded and equally the same everywhere. The hypothesis that "all places are alike" (Einstein's words) on large scales is today known as the cosmological principle.

The epic poem *On the Nature of the Universe* by Lucretius, in praise of the infinite atomist universe, was discovered in 1417 hidden away in a monastery. Probably Nicholas of Cusa was influenced by this work despite its atheistic tone.

Nicolaus Copernicus, an alumnus of the universities of Bologna and Padua, and a canon in the cathedral town of Frauenburg, was aware of the dormant heliocentric system of the ancient world and had the bright idea that a Sun-centered system might possess simpler movements and greater harmony. For years the canon labored on the computation of heliocentric orbits. His problem was to fit the movements of the planets to observations made from the Earth, which itself moved as one of the planets. At last he succeeded in showing that his new system of epicycles worked as well as the Ptolemaic system. He had the temerity to believe in the reality of the heliocentric system and did not think it just a convenient model devised to make calculations simpler, as piety demanded. In his work *Revolutions of the Celestial Orbs*, published shortly before his death in 1543, Copernicus wrote, "Why then do we hesitate to allow the Earth the mobility natural to its spherical shape, instead of supposing that the whole universe ... is in rotation?" His reasons for supposing that the Earth rotates were much the same as those offered earlier by Oresme. From a rotating Earth it was a short step to the idea of an Earth revolving around the Sun: "We therefore assert that the center of the Earth, carrying the Moon's orbit with it, passes in a great orbit among the other planets in an annual revolution around the Sun; that the Sun is the center of the universe, and that whereas the Sun is at rest, any apparent motion of the Sun can be better explained by motion of the Earth."

In the same year Andreas Vesalius published his great work *On the Structure of the Human Body*, of equal merit and possibly even greater originality, but of less cosmic significance.

Thirty-three years later the English astronomer Thomas Digges published a popular account of the Copernican system and discarded the outer sphere of fixed stars, which Copernicus had retained. Digges peeled away the outer sphere and distributed the stars throughout endless space: "Of which lightes celestiall, it is to bee thoughte that we onely beholde sutch as are in the inferioure partes of the same orbe, and as they are hygher, so seeme they of lesse and lesser quantity, even tyll our sighte beinge not able farder to reache or conceyve, the greatest part rest by reason of their wonderfull distance invisible unto

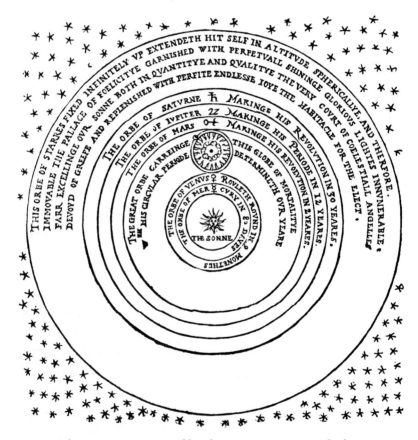

The universe as conceived by Thomas Digges in 1576. The finite Copernican system has become an infinite system in which the stars are dispersed through endless space.

us." The stars "devoyd of greefe" and serving as the "habitacle of the elect" occupied the "gloriouse court of ye great god." The *primum mobile* had gone, but the empyrean lingered on, and Edmund Spenser in his *Hymn of Heavenly Beauty* rejoiced in the new vision:

> Far above these heavens which here we see
> Be others far exceeding these in light,
> Not bounded nor corrupt, as these same be,
> But infinite in largeness and in height.

The bishop, cardinal, and canon had done their work well.

The fiery Dominican monk Giordano Bruno took to the warpath as the champion of the infinite universe. In his writings and travels he propagated the message of an endless universe teeming with countless worlds, each an abode of life with its own incarnation, revelation, and redemption. In 1584 he wrote, "Thus is the excellence of God magnified and the greatness of his kingdom made manifest; he is glorified not in one, but in countless suns; not in a single earth, but in a thousand, I say, in an infinity of worlds."

Condemned as a "malevolent witch," Bruno spent his last seven years shackled in an ecclesiastical prison, was tortured, and in 1600, with his tongue skewered between his cheeks, according to one witness, he was burned at the stake in Rome.

*   *   *

Tycho Brahe, a Danish nobleman of the second half of the sixteenth century, turned to astronomy and made observations of the utmost precision possible before the invention of the telescope. In 1572, a bright light flared in the constellation of Cassiopeia and slowly waned in the following months. The astonished Tycho, after careful observations, found it was indeed a star at great distance. The firmament and its myriad stars was therefore not as perfect and unchanging as everybody supposed. Five years later a great comet appeared, and again by careful observations Tycho showed that it could not be a fiery atmospheric phenomenon of the sublunar sphere, for it passed among the planets far beyond the sphere of the Moon. He came to the conclusion that the motions of comets proved that the crystalline planetary spheres of Aristotle could not exist.

Tycho agreed with Copernicus that the planets revolve around the Sun, but he compromised by constructing his own system in which the Earth remained stationary at the center of the universe and the Sun with its retinue of encircling planets revolved around the Earth.

When news reached Italy of the invention of the telescope in Holland, Galileo Galilei constructed his own telescope and was soon using it for astronomical observations. In 1610 he published a small book entitled *The Starry Message* in which he summarized his discoveries. On the title page he announced:

> The Starry Message – unfolding great and marvelous sights, and
> proposing them to the attention of everyone, but especially
> philosophers and astronomers, being such as have been observed
> by Galileo Galilei, a gentleman of Florence, professor of
> mathematics in the University of Padua, with the aid of a
> telescope lately invented by him, respecting the Moon's surface,
> an innumerable number of fixed stars, the Milky Way, and
> nebulous stars, but especially respecting the four planets that
> revolve around the planet Jupiter at different distances and in
> different periodic times, with amazing velocity, and which, after
> remaining unknown to everyone up to this day, the author
> recently discovered.

Galileo believed in the Copernican heliocentric system, and what he saw through his telescope – mountains on the Moon, many hitherto unobserved stars, and the satellites of Jupiter – strengthened his conviction. By observing sunspots he found that the Sun rotates. He noticed that the changing phases of Venus resemble those of the Moon, thus proving that Venus revolves around the Sun. He declared in his forthright manner that to the observant eye and the unprejudiced mind it was obvious that the Earth revolves around the Sun in company with all the other planets. But few of his contemporaries were able and willing to agree with him.

Galileo brought together the strands of medieval thought concerning space, time, and motion, and repeated the argument that a state of rest is relative and no more natural than a state of uniform motion. Using the idea of impetus (still not clearly defined), he investigated the paths of projectiles; by rolling balls down an inclined surface he confirmed that falling bodies accelerate at a rate independent of their weight; he showed that a pendulum swings with a period depending on its length but not the weight of its bob. At the age of sixty-eight, in 1632, Galileo published his masterpiece, *Dialogue Concerning the Two Chief World Systems*, in which he contrasted the Ptolemaic and Copernican systems and poured scorn on the physics of Aristotle and the astronomy of Ptolemy. For years his most hostile critics had been academics steeped in Aristotelian doctrine; the clearly heretical character of the *Dialogue* enabled his opponents to deliver him into the hands of the Church. Before the steely sauvity of the inquisitors, and confronted in old age with the prospect of torture, he recanted and abjured the heliocentric system.

Galileo inherited a rich legacy of medieval science, which he analyzed, synthesized, and popularized. He did not perform all the experiments attributed to him; some were performed by persons whom Galileo failed to acknowledge, and others were imaginary (thought experiments) made possible by his intuitive grasp of scientific principles. He failed, however, to apply mechanical principles to the celestial motions and failed to appreciate Kepler's work on this subject.

Living in Germany during the time of Galileo, Johannes Kepler also believed in the Copernican system, and with enthusiasm he adopted the new philosophy, calling all in doubt. Kepler agreed with Tycho, "there are no solid spheres," for how could the spheres exist and not be shattered by the passage of comets? He rejected the idea of an infinite universe that existed without center and edge. The very notion, is terrifying, and "one finds oneself wandering in this immensity in which are denied limits and center, and therefore also all determinate places."

Kepler inherited Tycho's astronomical records and for years strived to explain the movements of the planets within the framework of a Sun-centered system. From this work emerged his three famous laws of planetary motion that served as stepping stones to the Newtonian mechanistic universe. The first law, the best known and the only one we need mention, states that the planets move in elliptical orbits about the Sun. After two thousand years astronomy was at last free of its epicyclic fixation.

*   *   *

René Descartes, in the first half of the seventeenth century, helped to clarify the still murky notions of matter in motion. He enunciated laws that foreshadowed the work of Newton. By uniting algebra and geometry, thus forming a new branch of mathematics, he forged an essential tool for the mathematization of the mechanistic universe.

In many subjects Descartes kept to the beaten track. He echoed Aristotle: "a vacuum is repugnant to reason," for space by itself is nothing. Where there is no matter there can be no space. Descartes rejected the atomic theory by arguing that matter is essentially infinitely divisible. If matter were atomic, nothing would exist between the atoms, and a void would surround each atom. But a void is contrary to reason and therefore atoms cannot exist. Matter exists everywhere, and there is no space where there is no matter. Even in what is said to be a vacuum, matter is spread out thinly and continuously.

Descartes condemned astrology, the science of long-range astral forces, and believed that bodies influence one another only by direct contact. His guiding principle – action by direct contact – swept away all astrological hocus-pocus. The far-fetched and arcane notion of forces and other influences acting at a distance across space was too preposterous to be taken seriously. He pictured the Solar System as a great whirlpool of tenuous fluid in which the planets and satellites were entrained in vortical motions, much like floating leaves twirling on the surface of eddying water.

Natural philosophers of the seventeenth century, particularly in France, Germany, and the Low Countries, took their cue from Descartes (and were thus Cartesians) and would have nothing to do with occult forces of a long-range character. Up until the early decades of the eighteenth century the Cartesians believed that everything was pushed by pressure or pulled by other forces in direct contact, and weird and wonderful were the mechanisms they devised to explain the rise and fall of the tides.

Meanwhile, across the Channel dividing England from France, the magi of a new age were pondering on these matters. Later they would be accused by indignant Cartesians of creating a universe controlled by astrological forces. A universe nonetheless that would eventually enable men to land on the Moon.

*   *   *

To educated persons of the Greco-Roman world and the high Middle Ages the weight of a thing was nothing more than its desire to descend to the center of the Earth. The burden of heaviness, called gravity, was the natural penalty of earthly existence. Levity, the opposite of gravity, was the innate desire of less-ponderable things, such as fire, to ascend to more airy and etherial altitudes. The opposing desires of gravity and levity caused all things to seek their proper station in the sublunar realm.

Untutored barbarians experienced great difficulty with the idea of a spherical Earth and could not understand why people on the other side of the globe, standing upside down, did not fall off the surface. But educated persons trained in the rudiments of Greek science had no such difficulty. Wherever a person stood on the surface of the globe the downward direction was toward the center, and gravity was the worldwide desire of all ponderable things to reach that center. The bishop, cardinal, and canon had much to say on the subject of gravity.

Oresme dismissed the notion of levity as superfluous. All things have weight, he said, and whatever rises, such as warm air from a candle flame, is pushed or forced to rise by the descent of what has stronger gravity, such as cold air. Nicholas of Cusa, pursuing Oresme's ideas, surmised that other worlds – created by the infinite power and wisdom of the supreme being – had also centers to which their various parts tended to gravitate. Copernicus in his *Revolutions* expressed the medieval view:

> I think that gravity is nothing more but a certain natural appetition which the Architect of all things has implanted in the individual parts in order that they may unite to attain unity and wholeness by adopting a spherical form. It must be assumed that this property is found even in the Sun, the Moon, and other planets in such a way that their observed, unchangeable, spherical form is assured.

Gravity until the time of William Gilbert was an innate desire urging each thing to move downward.

Etheric attractions and repulsions pervaded the medieval universe, reaching across the celestial spheres, eliciting responses from angelic, daemonic, and human souls. The magnetism of lodestones was a convincing illustration of the existence of intangible astral and esoteric forces. Magnets were carried as charms to ward off the mischief of demons and witches, and animal magnetism and magnetic power are terms still in use.

William Gilbert, an Elizabethan physician and scientist, became widely known for his learned book on magnetism, published in 1600, in which he showed that the Earth is a huge magnet having north and south magnetic poles. He dismissed many superstitions about magnetism; for example, he showed that garlic could not demagnetize, and assured mariners that their garlic-perfumed breath would not enfeeble their compasses. He coined the word electricity, and conjectured that the planets were attracted to the Sun by an intangible force of a magnetic nature.

Gilbert promoted the atomist concept of an infinite universe strewn with countless inhabited worlds. The English at this time were moderately tolerant of their intellectuals, who had the sense to contribute to this happy state of affairs by bowing to the sensitivities of theologians and acknowledging the handiwork of God in every subject touched upon. Gilbert was the last person to scoff openly at contemporary religious beliefs, and yet his cosmology was as broad in scope as that of Bruno. Instead of being condemned as a heretic he was knighted by the queen and made court physician.

At a time when many natural philosophers in Europe had abandoned the medieval universe, their English colleagues held to the picture of a universe suffused with spiritual emanations. We need only read the work of Henry More, mentor to Newton, to realize at Cambridge that space was permeated with spirit. The Cartesian belief that by itself space could not exist without the support of matter was unacceptable to the English natural philosophers – the Newtonians – for a very good reason. The Newtonian universe, now in the making, consisted of the old atomist universe permeated by medieval spirit. Where there was no matter, spirit alone supported the extension of space. The Aristotelian and Cartesian belief that space could not exist without matter denied the omnipresence of spirit. To say that God was infinite and everywhere and then attribute the property of extension solely to matter was illogical.

In *Immortality of the Soul* Henry More in 1662 went so far as to define the nature of spirit:

> I will define therefore spirit in general thus: a substance penetrable and indiscerpible [invisible]. The fitness of this definition will be better understood, if we divide substance in general into these first kinds, viz. body and spirit, and then define body as a substance impenetrable and discerpible. Whence the contrary kind is fitly defined as a substance penetrable and indiscerpible.

Penetrable and invisible spirit endowed space with an innate reality independent of the presence of impenetrable and visible matter. The fusion of the atomist and medieval universes was of vital importance in the development of the Newtonian mechanistic universe of atoms and universal gravity. Scientists feel embarrassed on learning that the noble Newton dabbled in mysticism and often fail to understand that without mysticism there might not have been a Newtonian universe. We always write history in such a way that its events are rational to us rather than the players. The divestiture of space of spirit was the work not of the Newtonians but of the deists who followed in the eighteenth century.

We do not know for certain who was the first to identify gravity at the Earth's surface – a general appetition for wholeness – with the force reaching out and pulling on the Moon, and more generally, with the force reaching out from the Sun and pulling on the planets. (Possibly Robert Hooke was the first.) An inner desire to reach the center of the Earth had to be sublimated into a universal force issuing from each body and attracting all other bodies. Without doubt, the spiritual emanations of the medieval universe, retained and refined by the Newtonians, inspired the flight of fancy that linked earthly appetition with universal gravity.

*   *   *

Robert Boyle, Edmund Halley, Robert Hooke, Isaac Newton, and Christopher Wren rank among the principal Newtonians. Consider Wren, at the age of twenty-five, delivering a lecture at Gresham College in 1657 and in the name of Seneca prophesying:

> A time would come when men should be able to stretch out their eyes, as snails do, and extend them fifty feet in length; by which means they should be able to discover two thousand times as many stars as we can; and find the galaxy to be myriads of them; and every nebulous star appearing as if it were the firmament of some other world, at an incomprehensible distance, bury'd in the vast abyss of intermundious vacuum.

We remember Wren as an architect for his design of St. Paul's

Cathedral, built after the Fire of London in 1666, and not for his mathematics and imaginative contributions to science.

Consider Halley, famed for his discovery of the cyclic return of Halley's comet, which was last seen in 1986 and will return in 2061. Only three decades after the death of Galileo the pace of scientific development was breathtaking; Halley had established an observatory in the Southern Hemisphere, discovered the movements of stars, detected for the first time a globular cluster of stars, and plotted the paths of comets.

Consider the indefatigable Hooke, a thorn in the side of Newton, who for a time was Curator of Experiments at the Royal Society. He performed many ingenious experiments, made suggestions concerning practical devices such as steam engines, and pioneered the microscope (introducing the word *cell* into biology); he developed a theory of earthquakes, a wave theory of light, and opened up a vision of universal gravity. Bodies move in straight lines when free of applied forces, as shown by Descartes, said Hooke, and bodies in circular motion are subject to a centrifugal force, as is well known to all. The planets do not move in straight lines but in circular orbits about the Sun; hence the planets experience a centrifugal force urging them away from the Sun and must be continually pulled back toward the Sun by the force that commonly is called gravity. In 1670 Hooke explained his system of the world:

> That all celestial bodies whatsoever have an attraction or gravitating power to their own centres, whereby they attract not only their own parts, and keep them from flying from them, as we may observe the Earth to do, but that they do also attract all other celestial bodies that are within the sphere of their activity, and consequently that not only the Sun and Moon have an influence upon the body and motion of the Earth, and the Earth upon them, but that Mercury, Venus, Mars, Jupiter, and Saturn also, by their attractive powers, have a considerable influence upon its motions, as in the same manner the corresponding attractive power of the

Earth hath a considerable influence upon every one of their motions also.

Hooke's "System of the World," which would lead, he said, to "the true perfection of astronomy," is the first statement on record concerning universal long-range gravity. By 1679, Hooke knew from Christiaan Huygen's mathematical treatment of centrifugal force and Kepler's third law of planetary motion that the Sun's gravity weakens as the inverse square of distance from the Sun. He lacked, however, the mathematical ability to convert his descriptive account into a system of precise laws.

* * *

A new epidemic plague in 1665 caused Isaac Newton at the age of twenty-three (and twenty-three years after the death of Galileo) to leave Cambridge and spend two years on his mother's farm at Woolsthorpe in Lincolnshire. During this period he investigated the properties of light, invented calculus, developed other mathematical subjects, and came to the conclusion that gravity reaches out and controls the motion of the Moon around the Earth and of the planets around the Sun. Years later, looking back on this period, Newton said, "In those years I was in the prime of my age for invention, and minded mathematics and philosophy more than at any time since."

Natural philosophers had assembled a host of thoughts and discoveries; it was the genius of Newton reflecting deeply for many years that synthesized these thoughts and discoveries into an organized whole. He enunciated the laws of motion in clear and precise form, taking into account the reciprocal actions of forces, and gave the theory of universal gravity a secure foundation. His work, *Mathematical Principles of Natural Philosophy* (written in Latin and known as *Principia*) was published in 1687. From basic principles and definitions Newton developed various propositions and then explained mathematically the elliptical orbits of planets, the twice-daily tides on Earth due to the attraction of the Moon and Sun, the equatorial bulge of the Earth owing to its rotation, and so on, until it seemed that all terrestrial and celestial phenomena were governed by mathematical laws of motion in a universe where each part gravitationally influenced all other parts in a precise and deterministic manner. A universe created by God in which the heavenly bodies, separated by abysses of empty space, glided serenely according to supernal laws revealed to the mind of man; a universe, said Newton, "whose body nature is, and God the soul."

"Whence is it that Nature does nothing in vain and whence arises all the order and beauty in the world," wrote Newton in his *Opticks,* and a few lines later, "is not infinite space the sensory of a Being incorporeal, living, intelligent, omnipresent?"

*   *   *

Atomism, a theory of bygone ages (developed by the "most celebrated philosophers of Greece and Phaenicia," said Newton), did not enter the mainstream of science until the seventeenth century. The liberal-minded Newtonians converted the old atheistic theory into a new theistic theory of the universe. The Cartesians refused to accept the atomic theory of matter because it required that bits of matter (atoms) were separated from one another by spaces empty of matter (voids), and voids, as the Cartesians knew, could not exist. But the Newtonians, armed with the idea of space existing naturally of its own accord by virtue of pervading spirit, had no fear of voids and were keen on the atomic theory, which Boyle applied enthusiastically with great effect to the study of gases. In his *Opticks* Newton wrote:

> It seems probable to me, that God in the beginning formed matter in solid, massy, hard, impenetrable, moveable particles, of such sizes and figures, and with such properties, and in such proportion to space, as most conduced to the end for which he formed them ... even so very hard, as never to wear or break in pieces; no ordinary power being able to divide what God himself made one in the first Creation.

Newton had much more to say on the atomicity of matter and even proposed that light is composed of small particles.

Analysis of the scientific method is rarely helpful when we try to understand those central and creative figures – for example, Pythagoras, Anaxagoras, Newton, Einstein – who shaped and directed the advance of science. History has freewheeling periods when science seems to consist of little more than reaping the benefits of preconceptions that Thomas Kuhn called paradigms. Puzzle-solving minds exploit the paradigms and diligent investigators explore their consequences. Then, lo! Along comes a person with an original style of thought who envisions a new scheme of thought, a new system of the world, and lays the foundations of yet another universe. How does this person wind an armature of coiled themes, and so artfully solder the connections that it generates sparks and lights up a new world of knowledge?

*　　*　　*

Richard Bentley, a young clergyman who later became a famous classics scholar of the eighteenth century, in 1692 gave a series of lectures entitled *A Confutation of Atheism,* in which he aimed to show how the marvels of the Newtonian universe gave proof of the existence of God. Before publishing the lectures he asked Newton for his comments, and in the ensuing correspondence, Newton remonstrated, "You sometimes speak of gravity as essential and inherent in matter. Pray do not ascribe that notion to me. For the cause of gravity is what I do not pretend to know and therefore would take more time to consider it." From various remarks, and his famous "I feign no hypotheses" in the second edition of *Principia,* it appears that Newton was unwilling to regard gravity as purely physical in nature. Though quantifiable, it was immaterial and mysterious; its existence furnished evidence of God's influence at work in the universe, and Newton shared Bentley's belief that gravity gave proof of God's existence.

The new theory of gravity was remarkable in another way: it reinforced the idea that the universe is centerless and edgeless, and therefore uniform and infinite. In one of his letters to Bentley, Newton wrote:

> It seems to me that if the matter of our sun and planets and all the matter of the universe were evenly scattered throughout all the heavens, and every particle had an innate gravity toward all the rest, and the whole space throughout which this matter was scattered was but finite, the matter on the outside of this space would, by its own gravity, tend towards all the matter on the inside, and by consequence, fall down into the middle of the whole space, and there compose one great spherical mass. But if the matter were evenly disposed throughout an infinite space, it could never convene into one mass; but some of it would convene into one mass and some into another, so as to make an infinite number of great masses, scattered at great distances from one another throughout all that infinite space. And thus might the sun and fixed stars be formed.

As Newton said, if the universe is of limited extent and has an edge, and therefore also a center of some sort, the attraction between the various parts would cause them "to fall down into the middle . . . and there compose one great spherical mass." But in a universe of unlimited extent, without edge and therefore center, there exists no preferred direction in which each part can be pulled. In the second edition of the *Principia* we find: "the fixed stars, being equally spread out in all parts of the heavens, cancel their mutual pulls by opposite attractions."

Bentley's task of confuting the atheists seemed not too difficult, either to him or Newton. Mysterious gravity in a world of inert matter was undeniable evidence of the handiwork of God. From the omnipresence of God in the medieval universe had come the infinity and uniformity of the Newtonian universe. Gravity, moreover, proved that the universe had no edge, was therefore infinite, and hence proved that God who had created the universe must be, by implication, infinite also.

Thus the Newtonians paid back their debt. Gravity, derivative from the notion of pervasive theistic spirit, clearly demonstrated that the universe was necessarily infinite and uniform, and in turn demonstrated that the Creator, who could not be less than the created work, was indeed also infinite and everywhere. No other proof of the existence and nature of God has ever matched the elegance and self-consistency of that offered by the Newtonians.

7 The Mechanistic Universe

The telescope, microscope, thermometer, barometer, precision clock, air pump, and other seventeenth-century inventions preceded the Age of Reason in the eighteenth century. The age of enlightened reason commenced with prophets proclaiming visions of a new universe: "I feel engulfed in the infinite immensity of spaces whereof I know nothing and which know nothing of me, I am terrified. . . . The eternal silence of these infinite spaces alarms me," said Blaise Pascal. And "behold a universe so immense that I am lost in it. I no longer know where I am. I am just nothing at all. Our world is terrifying in its insignificance," said Bernard de Fontenelle.

The mechanistic universe of the eighteenth century more than fulfilled the promise of the prophets. Outfitted with laws uniting the Earth and the heavens, with self-running celestial mechanistic systems distributed throughout endless space, with time ticking away regularly as in Huygens's precision pendulum clock, the mechanistic universe opened up the prospect of the human mind able at last to solve all the riddles of nature.

Lofty thoughts that formerly soared among the towers of the Eternal City descended to street level in an exhilarating new Earthly City. Pious otherworldly preoccupations transformed into practical worldly occupations. The reborn world was bright and young, free of the late medieval conviction that all was senile and exhausted. The rejuvenated human sciences, led by "lapsed Christians," surged forward, achieving reforms that nowadays we take for granted as characteristic of Western society. Law and justice, crime and punishment received fresh scrutiny in the light of reason; slavery practiced overseas by Europeans drew increasing condemnation; novel political credos inspired the framing of constitutions and bills of rights; the

Iyrical and dramatic arts gave birth to essayists and novelists. Also, deism supplanted theism among the enlightened.

* * *

With the mechanistic universe came the custom of referring to the "reign of law and order." When scientists speak of the laws of nature they have in mind such things as the laws of motion and inverse-square law of gravity, which reveal regularity and harmony. To the Newtonians, who were dyed-in-the-wool theists, the world was God's temple and the reign of law and order meant nothing less than governance by a supreme being. The Newtonians peppered their works with generous acknowledgments to the Almighty, the Supreme Wisdom, and the Ruler who had created the universe, was manifest in its wonder and glory, and was continually at work in the working of its laws and maintenance of its order.

As the Age of Reason unfolded in the eighteenth century, the need for the direct participation of a supreme being became less pressing. The Laws of God denoting theistic superintendence transformed into the Laws of Nature implicit in Nature itself; the theism of God's governance transformed into the deism of Nature's governance. God the First Cause, the Architect, the Author who had created the mechanistic universe was no longer employed as a maintenance mechanic. The universe of perfect law and sublime order was self-running and self-adjusting. The product was so good, as a current commercial says of a certain washing machine, it required no maintenance. The God of the deists withdrew from the self-running mechanistic universe into a background of abstract being and remained there as the indispensable architect of it all. "If God did not exist," said Voltaire the deist, "it would be necessary to invent him."

By emphasizing Nature and Nature's laws, the deists avoided direct reference to God and God's laws. The relation of human beings to Nature usurped the relation of human beings to God. The changeover from theism to deism opened up for exploration intellectual territory previously fenced off as holy ground. By searching for human-nature laws in the mechanistic universe, the renascent human sciences (such as sociology, anthropology, psychology, and economics) strived to emulate the successes of the natural sciences.

The Age of Reason in the eighteenth century brimmed with bright hopes, bubbled with utopian dreams, overflowed with youthful ebullience. Despite its ups and downs and its turbulent radicals, such as Thomas Paine demanding "life, liberty, and the pursuit of happiness," it was a period of moderate political equilibrium, of surging economic and industrial growth, of high finance and booming overseas trade. The harnessing of natural science to industry and the development of powerful steam engines gave birth to an industrial revolution that had its roots in the Middle Ages. Even the churches ceased to harry and torment heretics, and the last witches met their doom in Western Europe in the early decades of the eighteenth century. In the new universe with its new god such horrors were inhuman and ungodly.

Instead of being witch-crazy, the Europeans became project-crazy. Everyone, it seems, had a pet scheme: business projects, get-rich-quick projects, prison-reform projects, educate-the-poor projects, pave-the-roads projects, welfare projects, emancipation projects, get-rid-of-aristocrats projects, make-everyone-an-aristocrat projects, projects to establish overseas colonies, and hosts of others of every kind abounded, all championed with enthusiasm and optimism.

In the air was the heady realization that the ancient world had at last been overtaken in every field of human endeavor. And it was true. A many-sided civilization had emerged of altered mentality, of numerous minds striving individually and collectively, equipped with a universe of unlimited promise.

It was a cuckoo universe, enticing, seducing, then compelling worldwide adoption, usurping and throwing out the indigenous belief-systems of non-European nations.

*   *   *

Through the Age of Reason swept the notion of progress like a wind sweeping away the cobwebs and dust of ages. Society hummed with purposeful activity, and everywhere ran an awareness that things were going places. The Renaissance had only the dream, the hope of rivaling the wisdom and gracious living of classical antiquity. It had few thoughts on the possibility of progressing beyond the glories of the past.

Deistic historians traced the ascent of man, seeking to understand where human beings had come from in order to plot where they were going. To this end they reconstructed the whole of history and outfitted their virgin universe with a past as new as a bride's trousseau. Not just the old past with a few amendments and the latest chapter added, but a new past, in which the shuttle of the human story wove a fabric of novel design. A new universe, the deists discovered, needed a new history.

In the language of the revised history, religion translated into mythology and superstition, evil into ignorance, redemption into enlightenment, divine grace into human virtue, God into Nature, Providence into Progress, and last but not least, Judgment into Posterity. In the new history, the Garden of Eden was symbolic of the golden age of the Noble Savage, the Fall symbolic of the rise of organized religion and the tyranny of priests, and Judgment symbolic of the esteem of Posterity for prestigious works. The Elected – the Europeans – led out of Exile by the Goddess of Wisdom could look forward to a Promised Land overflowing with happiness, filled with the prospect of the perfectibility of man. Thomas Jefferson and Benjamin Franklin shared these views and thought that if only human beings could rise above their religious obsession with the sinfulness of human nature and be free of preoccupation with an afterlife, then all obstacles would vanish in the path of social progress.

The ancients regarded time as cyclic, with the endless return of golden ages alternating with dark ages. All that had happened yesterday and yesterday and yesterday would happen tomorrow and tomorrow and tomorrow. The king is dead, long live the king! The cycles of the Wheel of Time were nonprogressive. But the Persians had jumped off the treadmill of cyclic time with the idea of a single cycle that began yesterday with Creation and will end tomorrow with Judgment. The Wheel of Time became the River of Time with its progressive improvement and continual development.

According to the new universal history, God created the universe in the beginning, and thereafter time had ticked away as in a well-oiled clock. The Coming and the End were out of sight because the celestial machinery could never wear out. A bright future of unlimited progress stretched ahead in an unbroken expanse of time in which all the accomplishments thus far would fade into nothing compared with things to come. Instead of Judgment and the award of treasure in Heaven, the deists believed in Posterity and the award of treasured memory on Earth. We are the inheritors of the full consequences of this philosophy.

*   *   *

The heavenly rewards of the Eternal City have gone, replaced by honors, prizes, and awards in the Earthly City. When distinguished people die, obituaries and biographies are written, memorials erected, and commemorative prizes instituted. Thus is their memory preserved and they have life ever after.

It was once God who saw everything and rewarded good works. Now society judges, posterity rewards, and publicity not prayer is what truly matters. We no longer hold in our minds the belief that Someone is watching, who records our motives and deeds, and one day will judge us fairly and independently of what other people think. The watchdog who maintained the highest standards and could not be deceived has gone. Instead, each person strives for recognition by society, which will enshrine and preserve his or her memory. Publicity is all that counts, and the worst thing that can happen is to be ignored. Why do writers fill libraries with books, scientists seek to disturb the universe, architects reshape the landscape, celebrities promote themselves on talk shows, actors pretend what they are not in front of cameras, artists, historians, politicians and the rest try to make their mark in attention-grabbing works, and criminals gain gratification when their evil acts are trumpeted around the nation? The mainspring of this dynamism is the desire to gain immortality in the Earthly City.

* * *

The Age of Reason faltered with the romantic movement that rose in revolt against the savants and their mechanistic blueprints, with Blake's mysticism, and with Wordsworth's despairing cry "We murder to dissect." It certainly had reached a low ebb in the early nineteenth century when Thomas Arnold, historian and headmaster of Rugby, complained in a letter, "Rather than have Physical Science as the principal thing in my son's mind, I would have him think that the Sun went round the Earth and the Stars were mere spangles in a bright blue firmament." The enlightened honeymoon ended with the French Revolution, and in the wars that inundated Europe, human beings were as benighted as ever.

* * *

Across the vault of heaven stretches the wraithlike arch of the Milky Way – the *via lactea* – formed by the numerous stars and luminous gas clouds of our Galaxy. Thomas Wright of Durham believed that the Milky Way offered ample reason for glorifying the works of God.

Wright was a gadabout youth of sixteen years when Newton died in 1727 at the age of eighty-five. After settling down in marriage as a surveyor and a teacher of mathematics to "noble ladies," he turned his attention to the spectacle of the heavens. At first, he agreed with Newton that the stars were "promiscuously distributed through the mundane space." Later, he realized that the observed stars are not randomly scattered but appear to be arranged "in some regular order." He published his thoughts in 1750 in a book entitled, *An Original Theory or New Hypothesis of the Heavens, Founded on the Laws of Nature, and Solving by Mathematical Principles the*

General Phenomena of the Visible Creation; Particularly the Via Lactea.

Wright proposed two models of the Galaxy, and in the one of interest to us he arranged the stars in a disklike system rotating about a center. The Milky Way was the disk of stars seen from our position inside the disk. Being a diehard theist, he viewed the universe as an arena of theistic superintendance and proposed a Neoplatonic type of galactic center endowed with supernatural power. "At this center of creation," he wrote, "I would willingly introduce a primitive fountain, perpetually overflowing with divine grace, from whence all the laws of nature have their origin." A deist at this stage might have thrown the book aside in despair and missed Wright's most daring conjecture. Wright went on to suggest that the fuzzy and faint nebulae of the night sky are perhaps other creations or "abodes of the blessed," similar to our Milky Way, but very far away.

In the agile mind of Wright, the Newtonian universe of scattered stars had transformed into an endless vista of "abodes of the blessed," each a distant and gigantic system of stars like our Milky Way.

Immanuel Kant in the university town of Königsberg read a review of Wright's work. Four years later in 1755 he published his own book having the equally long title, *A Universal History and Theory of the Heavens; An Essay on the Construction and Mechanical Origin of the Whole Universe, Treated According to Newton's Principles.* In this work Kant constructed the most stupendous universal picture ever conceived.

According to Kant's version of Genesis, in the beginning was chaos, as proposed by "the ancient philosophers," and like those philosophers he assumed that the "first state of nature consisted of a universal diffusion of primitive matter, or of atoms of matter, as those philosophers have called them." Out of the vortical motions of chaos, under the influence of gravity, came stars that congregated to form the Milky Way. Kant then drew on Wright's suggestion. The distant nebulae seen in the night sky as small elliptical patches of fuzzy light were whirlpool milky ways at great distances, each similar to our Milky

Way. He went farther: not only were the stars clustered into milky ways (now called galaxies), each held together by its own gravity, but also the milky ways were themselves clustered together to form much larger systems (clusters of galaxies), each also held together by its own gravity. The clusters of milky ways, said Kant, were probably clustered to form much larger systems that in their turn were clustered to form yet vaster systems, and so on, in an endless progression of systems of increasing size, filling infinite space. Kant quoted Pope:

> Look around the world; behold the chain of Love
> Combining all below and all above,

and saw in the hierarchical universe a natural extension of the great chain of being. "The theory we have expounded opens up to us a view into the infinite field of creation, and furnishes an idea of the work of God which is in accordance with the infinity of the great Builder." Unlike Wright, Kant was a firm deist and believed the created universe so perfect that further theistic intervention was quite unnecessary: "God has put a secret art into the forces of nature so as to enable it to fashion itself out of chaos into a perfect world system."

William Herschel, born in Germany, lived in England from 1757 onward; aided by his sister Caroline he became the foremost astronomer in the Age of Reason. Both abandoned their musical careers because of a consuming interest in astronomy, and both devoted their lives to constructing telescopes and observing the heavens. Discovery of the planet Uranus brought fame to William. He was fond of pointing out that astronomy has much in common with botany. "The heavens are seen to resemble a luxuriant garden, which contains the greatest variety of productions." Stars evolve, have individual life histories, and at a glance we see them in their various stages of development. "Is it not the same thing, "he wrote in *The Construction of the Heavens*, "whether we live successively to witness the germination, blossoming, foliage, fecundity, withering, and corruption of a plant, or whether a vast number of specimens selected from every stage through which the plant passes in the course of its existence be brought at once to our view?" Because we cannot wait for an acorn to evolve into an oak tree, we can at least observe oaks in various stages of growth and piece together the life history of a typical oak tree. Similarly, astronomers cannot wait for a star to evolve and must piece together the life history of a star from the display of many stars in various stages of evolution.

The Herschels observed and cataloged numerous stars and nebulae and were undoubtedly the founders of modern astronomy. William, a true son of the Enlightenment had, not surprisingly, many simplistic beliefs. He thought it quite obvious that the Moon is inhabited and that beneath the bright atmosphere of the Sun lies possibly a cool surface also populated with living creatures.

When Napoleon Buonaparte became first consul of France in 1799, he appointed Pierre de Laplace, a mathematician, as minister of the interior, then fired him six weeks later for creating a bureaucratic nightmare by attempting to introduce "the spirit of infinitesimals into administration." On the occasion when Laplace presented to Napoleon a copy of his work *Celestial Mechanics*, Napoleon said, "You have written this huge work on the heavens without once mentioning the Author of the universe." To which Laplace replied, "Sire, I had no need of that hypothesis." In the sciences, henceforth, God was relegated to the role of designing the laws and molding the atomic parts, but was not required to appear in person.

*   *   *

In the seventeenth century, the Cartesians and Newtonians broke through the limits of medieval space, and in the boundless expanse of a new universe, human beings lost their privileged central location. But many believed that little had been actually lost, for human beings continued to figure prominently in the cosmic design and remained the most conspicuous members of the Great Chain of Being that linked them directly to the throne of God. Men and women still retained their central location in the much more important biological–spiritual universe.

In the nineteenth century, the savants of the mechanistic universe finally broke through the limits of medieval time, and the Beginning receded into the mists of unrecorded time. Genesis was controverted and the Great Chain crashed down. All physical forms of life, enmeshed in the cosmic gearwheels, became integral parts of the mechanistic synthesis. God's temple on Earth collapsed and all that seemed of highest value lay crushed in the ruins.

The personal philosophies of individuals, undermined by the latest cosmic revolution, became neurotic, even psychotic, and the consequences are apparent in the social pathology of our time. Many fled from an unbearable reality created by the rise of a new and frightening universe. They rallied to extremist groups, formed iconoclastic movements against this and down with that, reverted to antiquarian religions, flocked to political creeds that purported to give cosmic significance to life, grieved in counter-culture communities, or retreated into autistic worlds of secret knowledge.

It is hopeless trying to understand the history of the nineteenth century, with its fulminations from pulpit and platform, without realizing that people were struggling to save their imperiled fundamental beliefs that gave meaning and purpose to life on Earth; nor can we hope to understand the furor of the scenes enacted in the twentieth century unless we realize that societies were struggling to find new beliefs, often with dismal and tragic results.

In the nineteenth century, the entire mythic universe was on trial; at stake was the veracity of biblical records claiming that creation had occurred a few thousand years ago. The Mosaic chronology of scriptural records (derivative but deviating from the Babylonian chronology) sustained the deep-rooted belief that human beings were of paramount importance in the cosmic scheme and the universe had been created solely for them in the recent past. From the biblical records, Dante estimated that the creation of Adam occurred in 5198 B.C.; Kepler estimated that the creation of the world happened in 3877 B.C.; James Ussher, an Irish bishop, fixed the date of creation at 4004 B.C.; and the great Newton, in his *Chronology of Ancient Kingdoms Amended*, set the date at 3988 B.C.

Astronomy had regrettably been mechanized; however, little was lost for the heavens still proclaimed the glory of the Lord and conformed to providential law. But geology, probing into Earthly history, was quite another matter. Here was a domain of nature that lay outside natural law, in which miracles once had free play. At the gates of geology, said Thomas Huxley, "stood the thorny barrier with its comminatory notice – No Thoroughfare. By Order, Moses." Geologists and natural historians arguing against the brevity of life on Earth were heretics, if not downright atheists, seeking to disprove the truth of holy writ.

* * *

We must backtrack a little. Georges-Luis Leclerc de Buffon, keeper of the Jardin du Roi in the mid eighteenth century, proposed that a large comet had struck the Sun a glancing blow and that the ejected matter then condensed to form the planets of the Solar System. He estimated the Earth had taken 100,000 years to cool to its present temperature. To reconcile his calculation with Genesis, he suggested that each of the six days of creation was actually a period of very long duration, and "day" needed reinterpretation in the light of new knowledge. In his masterpiece of 1778, *The Epochs of Nature*, Buffon rolled back biblical time to a remote beginning and said that natural history is revealed in the archives of nature and must be regarded as a science on the same footing as astronomy. His seminal ideas concerning the antiquity of Earth and evolution of prehistoric life, his suggestions that coal deposits are the remains of prehistoric life and the Great Chain of Being a web of interconnecting links like chain mail provoked outrage on a scale that astonished him.

Denis Diderot, a French contemporary encyclopedist and a notorious free-thinking philosopher, argued that the work of Kant clearly showed that the age of the universe is not just hundreds of thousands of years but more probably "hundreds of millions of years." Nowadays, the age of the Solar System is known to be five billion years, and the age of the universe somewhere between ten and twenty billion years.

By the beginning of the nineteenth century it was impossible for natural historians to brush aside the accumulation of evidence from the study of fossils and rock strata. Compromise doctrines capable of accommodating the Mosaic chronology became the fashion. Georges Cuvier, a French naturalist and later chancellor of the University of Paris, argued that the Flood was a crucial event separating supernatural and natural history. Human beings were created just before the onset of natural history, and the soulless lifeforms of the fossil record lived in the antediluvian periods. Further elaborations soon became necessary. The globe had apparently been periodically visited by many catastrophes, such as life-destroying deluges, and newly created life had arisen in more advanced forms after each visitation had devastated the globe. Thus the control of natural and supernatural laws alternated and life was created anew in episodic acts of special creation.

The Scottish physician James Hutton had little patience with Mosaic chronology and proposed a uniformitarian doctrine. The geological record reveals, he said, continuous erosion and sedimentation acting over vast periods of time, so vast that there "is no vestige of a beginning – no prospect of an end." Declaring, "no powers are to be employed that are not natural to the globe, no actions to be admitted except those of which we know the principle," he laid in 1785 the foundations of modern geology.

Hutton's deistic picture, untrammeled by catastrophic acts of theistic intervention, was later adopted by the great Scottish geologist Charles Lyell in 1830 as the theme of his *Principles of Geology*. The upthrust and erosion of mountains, the sculpturing of landscapes, and the shaping of continents are the result of steady and natural processes acting over interminable ages, wrote Lyell. "Thus, although we are mere sojourners on the surface of the planet, chained to a point in space, enduring for a moment in time, the human mind is not only enabled to number worlds beyond the unassisted ken of mortal eye, but to trace the events of indefinite ages before the creation of our race." In like manner, he continued:

> We aspire in vain to assign the works of creation in space, whether we examine the starry heavens, or the world of minute animalcules which is revealed to us by the microscope. We are prepared therefore to find that in time also the confines of the universe lie beyond the reach of mortal ken. But in whatever direction we pursue our researches, whether in time or space, we discover everywhere the clear proof of a Creative Intelligence, and of His foresight, wisdom and power.

On the one hand, the catastrophists believed in periodic cataclysms followed by acts of special creation, of which the last act occurred only a few thousand years ago; on the other hand, the uniformitarians believed in a single created state of long ago, and natural laws have since held uninterrupted sway. Catastrophe versus uniformity sounds nowadays much like big bang versus steady state in modern cosmology, but the controversy that raged in the early decades of the nineteenth century was far more heated than the debate between big-bangers and steady-staters in mid twentieth century. The catastrophists lost in the nineteenth century but won in the twentieth century.

*   *   *

The notion of evolution was in the air affecting the climate of opinion. Most members of the public accepted social evolution as synonymous with progress. One had only to compare the lifestyles of civilized and uncivilized people to see that social evolution obviously occurred. Across the gap separating European and primitive cultures stretched a chain of progressive social evolution. But it was not social but organic evolution that caused all the trouble.

The century of evolution – the nineteenth century – began with Jean Baptiste de Lamarck, a French naturalist who popularized the word biology. He resuscitated the old idea that organic life evolves from rudimentary beginnings and advances as it adapts to the exigencies of the environment. The cardinal idea of Lamarckian evolution was that creatures evolve organically in response to their needs and desires, and therefore evolution is "self-directing." Common sense urges us to believe that skills and aptitudes acquired by parents are inherited by their offspring. We feel that progress of this kind is by far the most valuable asset that we can hand on to our descendants. Such was the thrust of Lamarck's argument. But common sense is wrong, and so was Lamarck, and in matters concerning evolution common sense often misleads. If you believe that by your study and athletics your children will be studious and athletic, you are dead wrong; at most they will benefit by your example.

Robert Chambers, one of two brothers who founded the famous *Chambers' Encyclopaedia*, was a persuasive writer who popularized the Larmarckian theme. He performed, in Darwin's words, valuable service "in removing prejudices, and in thus preparing the ground for the reception of analogous views." Robert Chambers' widely read book *Vestiges of the Natural History of Creation* was as sensational in its day as Charles Darwin's *Origin of Species* fifteen years later. Competition and the struggle for life dominated the whole period of prehuman history, explained Chambers, and "the adaptation of all plants and animals to their respective spheres of existence was as perfect in those earlier ages as it is still." The struggle to survive and the ensuing adaptation by plants and animals to the vicissitudes of the environment was in full accord with natural law. Theistic intervention was quite out of court:

> We have seen powerful evidence that the construction of this globe and its associates, and inferentially all other globes of space, was the result, not of any immediate or personal exertion on the part of the Deity, but of natural laws which are expressions of his will. What is to hinder our supposing that the organic creation is also a result of natural laws, which are in like manner an expression of his will?

Human beings still believed in their own special creation, and were not a part of the evolutionary scheme.

At about this time Herbert Spencer developed a comprehensive scheme of evolution embracing biology and sociology. Following Chambers, he advocated a principle of "law versus miracle" and argued that evolution is "not an accident but a necessity," and used the phrase "survival of the fittest." As with others of his day, Spencer had difficulty distinguishing between evolution (denoting change of any kind occurring over time) and progress (denoting improvement judged by value concepts). Evolution, said Spencer, moves "from an indefinite, incoherent homogeneity to a definite, coherent heterogeneity," and is the development of systems of greater differentiation combining greater integration. When we compare an amoeba with the human body, we see in the human body greater organic differentiation combined with an effective integration of its parts.

We may interpret Spencer's theory rather freely as follows. Let the degree of integration of a system – either an organism or a society – be represented by the symbol X, and its degree of differentiation by the symbol Y. According to Spencer the product XY increases with progress and is a measure of the excellence of the system. When X and Y are both large, the system has simultaneously a high degree of integration and a high degree of differentiation; in other words, the system is generalized and yet specialized in many ways. Thus in a society, if X and Y are both large we have social order and individual freedom; if X is small and Y large, then disorder and anarchy; and if X is large and Y small, then order and totalitarianism (note that fascism, communism, and socialism imply large X but not large Y).

In Western society we tend to stress the importance of organization and standardization, and X is undoubtedly now larger than in previous centuries. But individual freedom, measured by Y, has not increased to the same extent and may in recent decades actually be decreasing. Uniformity can easily be attained, as in time of war, when freedom to be different is sacrificed for the sake of greater integration. But a harmonious society in which individuals have freedom to be different is much more difficult to attain. Social progress means that integration and differentiation both increase, or at least their product increases, and not just one at the expense of the other. Spencer's argument on combined integration and differentiation as the hallmark of progress is intriguing, especially when in a reflective mood we gaze at the scurrying in an ant heap, or stand at a street corner and gaze at the scurrying of a city and wonder how large is X and how small is Y.

*   *   *

The evolutionary theory of natural selection was independently proposed in 1858 by Alfred Wallace and by Charles Darwin. A year later, in his great work, *On the Origin of Species*, Darwin published his thoughts and investigations of many years. It suffices here to say that the new theory of natural selection flowed from four basic streams of thought.

The first stream is that all organic life evolves naturally and not supernaturally.

The second stream concerns the antiquity of the Earth. Darwin was influenced by Lyell and wrote, "He who can read Sir Charles Lyell's great work on the *Principles of Geology*, which the future historian will recognize as having produced a revolution in natural science, and yet does not admit how vast have been the past periods of time, may at once close this volume."

The third stream, familiar to breeders, is the knowledge that members of a species are not all exactly alike but differ from one another in various ways.

The fourth stream is the realization that the growth of populations is checked by environmental limitations; this last stream had its source in the Reverend Thomas Malthus's *Essay on the Principle of Population*, written in 1798. It is a strange and interesting fact, observed Malthus, that the human population is held in check by war, disease, and premature death, and is thus prevented from overburdening the available natural resources of society and of the land.

From the confluence of these four streams came the mainstream of natural selection. Given natural laws and sufficient time, individual differences favoring survival and reproduction are shared increasingly among the members of an interbreeding population, and as a consequence, the species evolves. Though producing many twists and turns in response to the environment, sometimes resulting in bizarre lifeforms, the natural selection of individual variations by differential reproduction is as inexorable as any other law of nature.

Darwin did not understand the cause of individual variations within a species. We now know that the genetic coding in twin-stranded molecules of nucleotides determines organic structure; small variations (mutations) in the coding are responsible for the individual variations within a species and are the natural consequences of molecular rearrangements inevitable in the chemistry of complex systems.

Through the genetic coding in the cells of living creatures the past reaches out and confronts the present. Natural selection is a dynamic process because the lifeforms selected by past conditions now exist under present and often different conditions. It is a game of tag in which the past rarely catches up. The naturally selected become the selected unnatural; the fittest survive and become unfit and do not survive.

Natural selection is as blind as the law of gravity and does not guarantee progress of any desired kind. If stability is the prize, cockroaches and crocodiles are the winners. Evolution has nothing to do with progress. Each step is dictated by what survives and breeds, and whole species are blithely discarded that later, in the new environment, would have been superior in fitness to those who actually survived.

In some ways Lamarck was right. Human beings do have control over their evolution, but not quite in the way he thought. The environment plays the tune, life dances accordingly, and humans fiddle with the environment. The natural environment is fast vanishing and being replaced with a dense matrix of human beings and their artifacts. The biosphere, now in a precarious state and ransacked for the purpose of maximizing the number of human beings, is fast becoming catastrophically unstable. Modern Lamarckism, it seems, is as blind as natural selection; perhaps worse, for natural selection at least has produced the human species, whereas Lamarckism in its present form is only capable of presiding over its demise.

*   *   *

Evolution means "unrolling" or "the appearance in orderly succession of a long train of events." The Sun evolves. Hydrogen in its deep interior slowly burns into helium and in roughly five billion years the Sun will evolve into a red giant, then into a white dwarf star. Evolution in astronomy applies to slow secular change of equilibrium configurations, and to episodic transformations, as in novae and supernovae. Stellar evolution does not imply "progress" of any kind and no astronomer would dream of such an implication. Evolution, as used in astronomy, has a simple and straightforward meaning consistent with a mechanistic treatment in the physical universe.

In the biological sciences the word evolution unfortunately is saturated with value concepts. The notion of social progress in the eighteenth century carried over into the notion of organic progress (evolution) in the nineteenth century. Progress became evolution, and in so doing evolution retained much of the meaning of progress. Progress means improvement, a change to better things, from lower to higher levels, and involves value judgments that are empty of physical content. The notion that evolution generally is progressive is deeply embedded in the language of the biological sciences: things that evolve are things that generally advance and improve.

Owing to the aura of progress investing the notion of evolution, we use fittest, advantageous, and other terms that are saturated with value concepts. When we try to justify our value concepts we find ourselves trapped in circular argumentation. Individuals surviving are the fittest, but what are the fittest? Obviously, those that survive. Individuals having advantageous variations reproduce and flourish, and what are advantageous variations? Obviously, those that reproduce and flourish. Whenever a value judgment trespasses into the physical universe it chases its tail.

The seductive word evolution, haloed in the life sciences with the mystique of progress, has no place in the mechanistic processes of the physical universe. Biological evolution either takes place or does not take place in the physical universe; there can be no fudging with a halfway world that is neither one thing nor the other, not if we wish to be rational in the universe of our society. As I see it, the law of natural selection is a physical law and must be treated as such. My modest proposal is that the word evolution should be used only by astronomers who have retained its proper meaning; natural historians would confuse us less if they stuck to safe words such as change and alteration.

*   *   *

The Origin of Species closes with the sentence, "There is grandeur in this view of life, with its several powers, having been originally breathed into a few forms or into one; and that, while this planet has gone cycling on according to the fixed laws of gravity, from so simple a beginning endless forms most beautiful and most wonderful have been, and are being, evolved." The clockwork universe of natural laws, extended endlessly in space and to the limits of time, embraced all animate as well as inanimate things.

The chilly light of the mechanistic universe banished the last shadows of the mythic universe. In their awesome universe, shivering human beings try to reassure one another by praising its glory and wonder. Edwin Burtt in The Metaphysical Foundations of Modern Physics wrote,

> The world that people thought themselves living in – a world rich in color and sound, redolent with fragrance, filled with gladness, love and beauty, speaking everywhere of purposive harmony and creative ideals – was crowded now into minute corners in the brains of scattered organic beings. The really important world

outside was a world hard, cold, colorless, silent, and dead; a world of quantity, a world of mathematically computable motions in mechanical regularity. The world of qualities as immediately perceived by men became just a curious and quite minor effect of that infinite machine beyond.

We come at last to the twenty-first century. Adrift like shipwrecked mariners in a vast and personally meaningless mechanistic universe we are found clinging for life to the cosmic wreckage of ancient universes.

Part II   The Heart Divine

8   Dance of the Atoms and Waves

In everyday life we deal with things of sensible size – such as flower-pots and plants – and to understand these ordinary things we explore the worlds of the very small and very large. We delve into molecules and atoms and reach out to the stars and galaxies. Thus, we know that most atoms composing the Earth were made in stars that died long before the birth of the Sun.

This wide realm of nature, of things ranging in size from atoms to galaxies, is ruled not by the gods of antiquity, but by the laws of motion and the push and pull of electrical and gravitational forces. Electrical forces dominate on the scale of molecules and atoms, accounting for much of the intricacy of the very small; gravitational forces dominate on the scale of stars and galaxies, accounting for much of the intricacy of the very large. The exploration of this luxuriant garden of phenomena is in the care of physical sciences such as chemistry, biochemistry, geophysics, and astrophysics.

The great problems lying deep at the foundations of the physical universe are no longer found in this realm that stretches from atoms to galaxies. They are found in the outer realms of nature. When the scale of measurement decreases a hundred thousand times smaller than the size of atoms, and increases a hundred thousand times larger than the size of galaxies, we quit the lush middle realm and enter the outer realms. Here we discover the truly baffling. In the subatomic realm of the extremely small lies the enigmatic diversity of strange forces; in the cosmic realm of the extremely large lies the enigmatic unity of the physical universe.

We start in this chapter by exploring the atomic and subatomic realm.

*   *   *

Pierre Gassendi of the seventeenth century, a French professor of philosophy and mathematics who believed that happiness consists of the harmony of body and soul, sought to revive the dormant atomic theory. He stripped from the atomic doctrine of the ancient world its outspoken denial of the gods. The Newtonians adopted his revived atomic natural philosophy and wove it into their theistically created mechanistic universe. After two thousand years the atomic theory came in from the cold of being tainted with atheism and at last became respectable.

René Descartes and the Cartesians who followed in his footsteps would have nothing whatever to do with the atomic theory. They insisted that matter necessarily was infinitely divisible and atoms therefore could not exist. But Robert Boyle, alchemist of a new age, showed how the idea of atoms explained the properties of gases. The final blessing was given by Newton who said there are "agents in nature able to make the particles of bodies stick together by very strong attractions. And it is the business of experimental philosophy to find them out." Physicists today follow Newton's advice and devote considerable time and effort to the business of finding out these very strong attractions in the atomic and subatomic world.

The atomic theory entered chemistry in the early nineteenth century. John Dalton, who investigated color blindness and yet himself was color blind, popularized the Greek word atom, meaning uncuttable. By supposing that atoms of different elements have different weights, Dalton showed how the elements combine in fixed proportions to produce chemical compounds. He was the first to make the atomic theory quantitative. The discovery of the negatively charged electron in 1897 by Joseph Thomson and the positively charged atomic nucleus in 1911 by Ernest Rutherford launched the modern era of atomic physics. Niels Bohr in 1913 constructed a model of the atom in which the electrons move in orbits about the nucleus. A few years later Louis de Broglie, Erwin Schrödinger, Werner Heisenberg, and other eminent physicists laid the foundations of the quantum mechanical model of the atom. The quantum world of the atom has changed our view of the physical universe and transformed the society in which we live.

An atom consists of a small positively charged nucleus surrounded by a cloud of negatively charged electrons. Many persons may vaguely recall hearing such a statement while dozing in the classroom. Probably the teacher was not a poet. "When it comes to atoms," said Niels Bohr, "the language that must be used is the language of poetry." Students wake up when told about cells, cytoplasm, organelles, chromosomes, and the double helix, for here are things of immediate human significance. They hear music in the litany of life-science terminology but not in the jargon of lifeless electrons, protons, neutrons, and other creatures of the quantum world. It is a pity, for without atomic particles there could be no organic life, and in the impalpable and seemingly inconsequential entities of the quantum world one finds the true music and magic of nature.

*   *   *

We have taken the first big step: the atom consists of a heavyweight tiny nucleus surrounded by a cloud of lightweight electrons – a step accompanied by many misconceptions. It was once the custom to imagine the atom as a miniature solar system in which electrons encircled the nucleus like planets orbiting a star. This idea still persists in popular literature. But electrons do not move in clear-cut orbits like orbiting celestial bodies. Instead, they dance and the atom is a ballroom. They perform stately waltzes, weave curvaceous tangos, jitter in spasmodic quicksteps, and rock to frenetic rhythms. They are waves dancing to a choreography different for each kind of atom.

The old idea that subatomic particles are similar to billiard balls is out. Piet Hein in *Atomyriades* miscued when he wrote the lines:

> Nature, it seems is the popular name
> for milliards and milliards and milliards
> of particles playing their infinite game
> of billiards and billiards and billiards.

An elementary particle is not like a billiard ball. It is a vibrant mysterious world cunningly created.

An electron consists of waves. The waves – spreading out and interweaving wherever possible – account for the structure and behavior of atoms. They lace together arrays of atoms into molecular tapestries and create the rich and varied patterns of our world of plants and flowerpots.

How can a tiny electron behave like a widespread wave? We must face the fact, as much a fact as any we know in the physical world, that all subatomic particles, not only electrons, have a remarkable dual nature. Or rather, we ascribe to them a dual nature to gain an understanding. At one moment a particle is like the ripples on the surface of a pond, and at the next moment it is like a pebble on the floor of the pond. A particle is wavelike and corpuscular, and its dual nature is as perplexing as the duality of mind and matter. When we observe a particle it seems corpuscular, when we explain a particle it seems wavelike. The quantum world contains nothing that resembles our world of commonplace experience and we must not try to comprehend things in the vulgar fashion. In *The Character of Physical Law*, Richard Feynman remarked, "I think I can safely say that nobody understands quantum mechanics." He meant that nobody can understand it with ordinary common sense. The new territory is bizarre; tourists marvel, and physicists take up residence.

An electron as a wavelike entity is widespread over regions of space. These regions are sometimes small, sometimes large. An electron (or any other subatomic particle) fills all accessible space with its waves. It spreads everywhere and when we succeed in observing it something odd happens: it collapses and becomes a sort of corpuscular entity active within a small region, and we can then say, Ah! now we know roughly where it is.

Consider the following situation. An electron is shot at a target. It travels not like a corpuscle but as a wave and has all the properties of a wave. But when it reaches the target, it strikes it not as a wave spread out over the target, but as a corpuscle at a point on the target, and may emit a scintilla of light betraying where it has landed. We never know exactly the spot at which the electron will strike and we can only estimate from its wavelike behavior the chance – or probability – of where it will land in corpuscular form. The probability of where it will appear is proportional to the square of the amplitude of the wave. Where the wave is strongest is where the electron has the best chance of being observed.

There is nothing chancy or uncertain about the waves themselves. They are totally predictable, and evolve in various ways from state to state and travel from place to place in a manner fully and accurately determined by the equations of quantum mechanics. Yet only from its amplitude can we estimate the probability of where and in what way the wave will collapse and become an observed corpuscular event. This is the heart of the mystery. The waves of the electron extend wherever allowed; they evolve and propagate in various ways, and when the electron is finally observed, it is not a wave but a discrete entity that appears somewhere or other, but where is never exactly known. We know only the chance or probability of where and when and how it will appear and be seen.

We have a wraithlike quantum world of ghostly waves where all is fully determined and predictable. Yet when we translate it into our observed world of sensible things and their events, we are limited to the concept of chance and the language of probability.

A single excited atom is a wavelike system that evolves continuously and predictably in a comprehensible manner. To the observer it has a probability of decaying abruptly at any moment, but when is never exactly known. The deterministic precision of the quantum world contrasts with the fortuitous imprecision of the observed world. We are sometimes puzzled, though most of the time we are not bothered very much. Often, we deal with numerous atoms whose average behavior is predictable. Many atoms together behave in a continuous manner, and we can predict statistically how many will decay in each interval of time. The continuous and predictable behavior of many atoms in the observed world mimics the continuous and predictable behavior of a single atom in the wavelike quantum world.

Light – and in fact all radiation – behaves as either waves or particles. The particles of light are known as photons. A wave of light irradiates a target, such as a photographic plate; the light is not absorbed smoothly and uniformly over the whole plate, as one would expect, but in discrete packages of energy, as photons, at random points. If the intensity of the light is made weaker and weaker, eventually a stage is reached at which we can detect and count the arrival of individual photons. Again, this is very puzzling. The rippling wave is everywhere; we understand how it travels in space and how it is incident on the target, yet when we try to observe it, we detect discrete photons and know only the chance of where and when they will appear.

We must understand that the electrons in an atom vibrate with rhythmic modes and are excited into states of various energies. As the modes of excitation evolve, waves of light are emitted and absorbed. These waves travel in space, and on their arrival at the retina of a person's eye, other atoms absorb the incident waves. This vibrational give-and-take of atomic energy, in which atoms act both individually and collectively, and exchange quantized amounts of energy, accounts for our visible world.

In the quantum world of potentiality we calculate how atomic waves and light waves evolve, and thus we know in the observed world the probabilities with which atoms make transitions and photons are emitted and absorbed. The quantum world is deterministic and virtual, the observed world is statistical and actual. Poets have yet to catch up with the antics of the atomic and subatomic world and to rhapsodize on the most marvelous things of nature.

* * *

Almost everybody has heard of the uncertainty principle and knows that it has something to do with the atomic and subatomic worlds.

Many feel tempted to shrug it off as physicists' sorcery. The uncertainty principle is easily stated: the more precisely we know where a moving particle is now, the less precisely we know where it was in the past and will be in the future. We are given a trade-off in precision that comes as a result of the wavelike properties of the quantum world.

The particle is distributed as a wave, and this wave tells us more or less what can be observed: how the particle moves (its velocity or, rather, its momentum) and where the particle is located in space. Both items of information cannot be known simultaneously with unlimited precision; there is a trade-off in the precision with which each may be known. The more we know of one, the less we know of the other, and this is a fundamental property of the physical universe.

The uncertainty principle can be stated alternatively in terms of energy and time. The longer we measure the energy the more precise its value but the less precise the corresponding moment in time. Ordinary waves traveling on the surface of water or through the air have a similar form of uncertainty. In the particle world, energy and frequency are equivalent. The longer the waves are observed, the more certain becomes their frequency, but the less certain we know the frequency at a moment in time. When the measurement lasts over a short time, the observed frequency is less certain but the time more certain; when the measurement lasts over a long time, the observed frequency is more certain but the time less certain.

Atomic particles behave as waves, and their wavelike character accounts for the remarkable properties of the uncertainty principle. A particle is like the ambisphaenic snake: you cannot know precisely from where it comes, and

> Before it starts you never know
> To what position it will go.

In the quantum world nature is like a generous bank that lends out energy free of interest. But all loans must be repaid within a specified interval of time. The larger the amount borrowed, the shorter the interval of time it can be used before it must be returned. Everywhere energy is borrowed and repaid continually by all atomic and subatomic systems, and this greatly adds to the intricacy of the dance.

The energy borrowed multiplied by the loan period cannot be more than a certain maximum value determined by Max Planck's fundamental constant. Enough energy can be borrowed to create a particle for a brief moment. The energy loan is then withdrawn, the ledger balanced, and the particle vanishes. These short-lived ephemeral entities are called virtual particles; they live brief, ecstatic moments on borrowed energy and are the same as ordinary particles in all other respects. These will-o'-the-wisp creatures are of considerable importance in the makeup of our world; they influence the natural states of atomic systems and are responsible for gluing together the atomic nucleus.

There are a few incidental complications. Nature, for instance, does not lend electric charge, and a virtual electron must be escorted by its virtual antiparticle of opposite charge. The antiparticle of the electron is the positron that is similar to the electron but has a positive electric charge. The combined charges of the two particles is zero, and hence only energy is used to create both simultaneously as a pair. Spin is another property of the electron, and the positron created with it has opposite spin. Constantly, energy is borrowed to create not only electrons and positrons, but also all other kinds of particles and their antiparticles. The whole of space is flooded with a sea of seething virtual particles, all popping in and out of existence in mind-reeling numbers. Everywhere, at any moment, one million million million million million virtual electrons exist in a volume equal to that of a thimble.

This raises an obvious question: Why do we not see, hear, touch, taste, and smell them? The answer is that each must repay every bit of its borrowed energy and is not allowed to spend even an iota of the loan to make itself known in our world of packaged and bartered energy. On its return to limbo it leaves behind not a vestige of its borrowed energy.

On rare occasions virtual particles succeed in finding enough energy from somewhere to pay off their debt; then, accompanied by antiparticles, they are released from the debtors' prison and are free to enter the real world. Newborn particles of this kind are all around us, created by energetic cosmic rays that pour in from outer space, or created in high-energy accelerators used by physicists for the study of subatomic structure. This does not mean that the total number of particles is on the increase. When an electron and a positron cease to be virtual, the positron quickly annihilates with the electron or some other nearby electron, releasing energy, usually in the form of photons, and we are left with the same number of electrons and positrons as before.

To make all virtual particles real would require the utmost energy, vastly beyond what is available today in the physical universe. Yet long ago, in the very early universe, there existed sufficient energy, and the multitudinous virtual particles were real and had their moment of glory.

*   *   *

The nucleus of the atom is itself a dance of waves to a rhythm twenty octaves higher than the stately waltz of the ambient electrons. This compact central region of the atom contains protons and neutrons, each much heavier than an electron. The proton has a positive charge and the neutron has no electric charge. The nucleus also contains virtual particles, such as pions, which skip to and fro among the protons and neutrons and account for the strong forces that cement the nucleus together. The nucleus of the hydrogen atom is the simplest and lightest of all nuclei, and consists of only a single proton. The nuclei of other atoms have protons and neutrons in various combinations; for example, the helium nucleus has two protons and two neutrons, and the iron nucleus has twenty-six protons and thirty neutrons.

In the atomic world, energy exists in two forms – chemical and nuclear – and to understand this state of affairs we need not be atomic wizards.

Most chemical energy is released and absorbed when atoms are combined into molecules of various kinds. Chemical energy is absorbed when food is cooked, and the absorbed energy is used for rearranging the atoms in the carbohydrate and protein molecules. Whenever wood, coal, or oil is burned, some of the atoms are rearranged into new molecules, such as carbon dioxide, and at the same time chemical energy is released and appears in the form of heat. Our biosphere is an exceedingly complex system of molecules that continually interchange energy. Some of this energy we redirect for our own use, which is a fine idea when not overdone. Like fishing the seas, if you take too much, the system breaks down, and you are left with nothing.

Sunlight bathes the Earth with photons of just the right range in energy for stimulating the formation of organic molecules. The chemical energy stored in wood and in the fossil fuels of coal and oil came originally from sunlight, and sunlight derives from nuclear energy. Nuclear energy is released and absorbed whenever protons and neutrons are combined and rearranged into nuclei of different sorts.

The distinction between chemical and nuclear energy is this: rearrangement of atoms in molecules involves the release and absorption of chemical energy; rearrangement of protons and neutrons in atomic nuclei involves the release and absorption of nuclear energy. An important difference is that the amounts released and absorbed are about a million times greater in nuclear energy than chemical energy.

Through my window I see the Sun. That shining orb, poised in the sky, is a titanic nuclear reactor. It is a star, a globe of hot gas held together by gravity, which consists mostly of hydrogen. It is slowly converting hydrogen into helium. The Sun's surface has a temperature of 6000 degrees, and its center a temperature of about 10 million degrees. Because of the high interior temperature, hydrogen atoms are stripped of their electrons, and the hot gas consists mostly of free protons and electrons moving around independently at high speed. In 1924, Arthur

Eddington portrayed the scene in *The Internal Constitution of the Stars*:

> The inside of a star is a hurly-burly of atoms, electrons and aether waves. We have to call to aid the most recent discoveries of atomic physics to follow the intricacies of the dance. We started to explore the inside of a star; we soon find ourselves exploring the inside of an atom. Try to picture the tumult! Disheveled atoms tear along at 50 miles a second with only a few tatters left of their elaborate cloaks of electrons torn from them in the scrimmage. The lost electrons are speeding a hundred times faster to find new resting places. Look out! there is nearly a collision as an electron approaches an atomic nucleus; but putting on speed it sweeps around it in a sharp curve. A thousand narrow shaves happen to the electron in 10^{-10} [one ten-billionth] of a second; sometimes there is a side-slip at the curve, but the electron still goes on with decreased or increased energy. Then comes a worse slip than usual; the electron is fairly caught and attached to the atom. Barely has the atom arranged the new scalp on its girdle when a quantum of aether waves runs into it. With a great explosion the electron is off again for further adventures. Elsewhere two of the atoms are meeting full tilt and rebounding, with further disaster to their scanty remains of vesture.

At the time when Eddington wrote these words he did not know that the Sun consists mostly of hydrogen; his vivid picture, however, needs little alteration. In the same vein, he continued: "As we watch the scene we ask ourselves, Can this be the stately drama of stellar evolution? The knockabout comedy of atomic physics is not very considerate towards our aesthetic ideals; but it is all a question of time-scale. The motions of the electrons are as harmonious as those of the stars but in a different scale of space and time, and the music of the spheres is being played on a keyboard 50 octaves higher."

Luminous stars like the Sun radiate energy into space for billions of years and therefore must have a long-lasting internal source of energy. This source was unknown in Eddington's day, although nuclear energy of some kind was suspected.

* * *

Protons (the nuclei of hydrogen atoms) in the deep interior of the Sun move around in all directions at high speeds, continually encountering one another. Each collides about a million million times a second with other protons. But because protons are positively charged, they repel one another, and when any two rush to meet each other, they are pushed back and turned aside by their mutual repulsion. Protons in ordinary stars have little chance of ever coming very close together. In the whole of the Sun not a single proton has enough energy to penetrate the electric repulsion barriers that keep protons apart.

Our picture of protons flying about is misleading. It overlooks the wavelike interaction between protons. Behaving as waves, like light feebly penetrating through a dark window, they occasionally filter through the repulsion barriers that normally keep them separated. About once every second each proton in the central region of the Sun succeeds in making a wavelike penetration and comes face to face with another proton. In these fleeting face-to-face encounters each brings into play its strong nuclear force. If that were the end of the story, it would also be the end of us. In one second only there would occur an immense release of energy and the Sun would explode.

Life exists on Earth because protons are shy creatures. When brought face to face, they take a considerable time in deciding whether to like each other. Before their minds are made up they have moved apart and gone their separate ways. A similar thing happens to people in cities; they move about, encountering one another on the streets and in the subway, and sometimes a person meets another for a fleeting moment and feels a strong attraction. But in their movement and hurry they turn aside and go separate ways, perhaps never again to meet. An attitude of reserve between strangers prevents instant intimate friendship. Protons have an equivalent inhibition, and their shyness and inability to make instant friendships is due to what is called the weak interaction.

The problem is this: no nucleus exists consisting of two protons only. Protons repel each other too strongly to form a nucleus. But a nucleus exists that consists of one proton and one neutron; it is the deuteron, the atomic nucleus of heavy hydrogen. When two protons come close together and interact strongly, one of them has got to change into a neutron (by emitting a positron and a neutrino) so that both can be wedded into a deuteron. This switching of identity, of one proton changing into a neutron, while both are close together and strongly interacting, involves the weak interaction, which works extremely slowly. The probability of forming a deuteron during the brief encounter is extremely small. A proton in the central region of the Sun takes typically 10 billion years to unite with another to form a deuteron. When this happens there is a tumultuous honeymoon and a release of nuclear energy.

Once a deuteron has formed, it combines in approximately 100,000 years with another deuteron to form a helium nucleus and further energy is released. The nuclear energy unlocked by the conversion of hydrogen into helium maintains the Sun as a luminous body and supplies the energy radiated from its surface as sunlight. The transmutation of hydrogen into helium is a slow and continuous process, and after about ten billion years the hydrogen in the center of the Sun will be at last exhausted. The lifetime of hydrogen in the center of the Sun is roughly the luminous lifetime of the Sun as an ordinary star. When the Sun has consumed its hydrogen, which will occur in about five billion years time (it is already five billion years old), it will swell into a red giant, then quickly turn into a white dwarf, "palely loitering" in the skies, with no further supply of nuclear energy.

Stars more massive than the Sun, after burning their hydrogen, become luminous stellar giants of even higher temperature and have access to further supplies of nuclear energy by burning helium into heavier elements, such as carbon and oxygen.

*   *   *

The biggest and nearest nuclear reactor in this part of the Galaxy is our genial lord and master the Sun, whose beneficent radiation derives from the release of nuclear energy. A technological dream of the modern age is to discover a way of burning hydrogen into helium, as in the Sun, and to release energy on Earth in a controlled and steady fashion.

Ten billion years is much too long a time to wait, and fortunately the slow weak-interaction courtship between protons can be avoided. Heavy hydrogen, whose atomic nuclei are in the form of deuterons, is moderately abundant and enough exists in the oceans to meet the energy needs of humans for millions of years to come. All we need to do is heat the heavy hydrogen to a temperature of about 100 million degrees; it will then burn to helium, and useful energy will be released without the long delay caused by the weak interaction. But so far we have not found how to do this in a way that liberates energy steadily and not explosively. We already know, heaven help us, how to ignite heavy hydrogen explosively in the hydrogen bomb. But how to do it in a controlled manner for the benefit and not the ruin of mankind still eludes us, despite the sustained efforts of many scientists over the last few decades.

Nuclear energy is obtained either by bringing together very light nuclei (fusion) or by breaking up very heavy nuclei (fission). The nuclear reactors we have at present do not obtain their energy from the *fusion* of hydrogen into helium, as in stars, but from the *fission* of uranium and plutonium into nuclei of lighter weight. From fission we get the heat that generates electricity and the plutonium of nuclear weapons. The nuclei of atoms have become the subject of alarming news. The fission method is messy; its ashes remain unavoidably radioactive for long periods of time, and we have as yet no foolproof method of disposing of them. "I fear the Greeks even when they are bearing gifts," said Virgil. The gift of atoms by the Greeks has brought us hazardous radioactive wastes and turned our world into a nightmare of nuclear weapons.

That sublime product of nuclear energy, the sunlight incident on the Earth's surface, is more than sufficient for our energy needs, if only we knew how to use sunbeams in heavy industry for feeding blast furnaces, steel mills, electric power stations, and the needs of transportation. The power we at present derive from sunlight and from geophysical sources would hardly suffice to maintain a medieval standard of living for our vast twentieth century human population.

*   *   *

We try to explain our world of plants and flowerpots – of ordinary and sensible things – by delving into atoms and reaching out to the stars. Yet in this quest we find not the simplicity we seek, but utmost complexity that itself has urgent need of explanation.

The rich diversity of our environment breaks down into an assortment of millions of different kinds of molecules, which themselves break down into an assortment of less than a hundred different atoms. At first glance, the atoms decompose into three different particles – electrons, neutrons, protons – and it must be admitted that thus far we have achieved considerable simplification in nature. But now, as we delve deeper, seeking to understand more, there opens up a subterranean world of depthless mystery.

Electrons, protons, and neutrons are only three of the numerous kinds of particles now known. The electron is the familiar example of the lightweight subatomic particles called leptons. At present there are six leptons: the electron and its neutrino, the muon and its neutrino, the tauon and its neutrino, and each has its antiparticle, making twelve leptons altogether. The proton and neutron are the familiar examples of the heavyweight subatomic particles called hadrons. The hadrons divide into two families – baryons and mesons – and hundreds of different specimens of both kinds have been found.

A few decades ago the hadrons were thought to be among the ultimate constituents of the physical world. Now we attribute their properties to the existence of more fundamental subatomic particles of a yet deeper realm, which combine in various ways to form hadrons.

These strange new creatures, called quarks by Murray Gell-Mann (from James Joyce's "Three quarks for Muster Mark!"), interact among one another with quixotic forces that fail to get weaker as their separating distances increase. When you try to tear quarks apart from one another, increasing their separating distances, their attractions remain strong, and the work performed in the attempt creates new quark combinations. Trying to separate two quarks is like trying to isolate the ends of a piece of string; if you pull hard enough, the string breaks, and you are left with two new ends. Trying to isolate the ends creates new ends, and trying to isolate the quarks creates new quarks. Quarks exist together in groups of two (forming mesons) and in groups of three (forming baryons) and cannot be observed as isolated entities.

At the moment of writing there are six leptons; also six quarks (distinguished by the names up, down, charm, strange, top, and bottom), and each quark is dressed in one of three colors. When we add them all together with their antiparticles, and include the eight gluons that mediate between the quarks, we get fifty-six distinctly different subatomic particles. To this sum must be added the photon of the electromagnetic field, the elusive graviton of the gravitational field, the particles mediating in the weak interaction, and yet others even more exotic.

We began by seeking to understand things of sensible size, such as plants and flowerpots. We delved down, through the molecular realm with its DNA and other elaborate structures, to the atomic realm with its apparent order and simplicity. Beneath the atomic realm has opened up a subatomic world of startling complexity that teems with particles of many kinds. For all we know this might not be the end of the search. Deeper realms may exist, consisting of even more exotic particles of even lusher variety. Although we tell ourselves that we are at last uncovering the ultimate secrets of nature, some of us have moments of misgiving. The secrets of the subatomic world seem more puzzling than the comparatively sensible atoms they purport to explain. Are we nearer to answering John Hall, the seventeenth-century poet?

If that this thing we call the world
By chance on atoms was begot
Which though in ceaseless motion whirled
Yet weary not
How doth it prove
Thou art so fair and I in love?

9   Fabric of Space and Time

We take space and time for granted. Normally they do not trouble us, yet whenever we think about them we become puzzled.

Space seems simple enough. Here it is, all around us, stretching away and spanning everything in the external world. We are surprised when told that people in other cultures have different ways of regarding space. What is there about it that can possibly be different? Edward Hall in *The Hidden Dimension* says, "there is no alternative to accepting the fact that people reared in different cultures live in different sensory worlds" – in other worlds of space. It seems that the Arabs, Japanese, Hopi, and the people of many other cultures have different modes of expression concerning arrangements and relations in space; they live in different mental worlds – in other worlds of space.

Time is much more puzzling. Here it is in our imagination, stretching away, spanning everything in the past, present, and future. But unlike space it is not all around us and directly accessible. We experience time within ourselves, it seems, and cannot perceive it directly in the external world. Those intervals of minutes and hours on the face of a clock are actually intervals of space. A second cannot be displayed directly in pure form in the external world in the same way as a centimeter. This lack of objectivity about time greatly puzzled Robert Hooke in the seventeenth century: "I would query by what sense it is we come to be informed of time." Space is out there and apparently objective, yet time is in here and apparently subjective. That other cultures have different ways of regarding time is not surprising. The Australian Aborigines believe in a dreamtime where the past with its ancestral figures coexists with the present.

Many ancient cultures believed in the Wheel of Time, in the eternal return of the same pattern of events: the Sun rising and setting, the Moon waxing and waning, the seasons coming and going, the king dying and yet living, birth and death alternating in repeated incarnations, nations triumphing over nations, catastrophe endlessly following catastrophe, wheels turning within wheels, cycles enfolding cycles, *yuga* following *yuga*, *maha yuga* following *maha yuga*, with the Days of Brahma numbered though seemingly endless; and the gods, creating and destroying worlds, themselves doomed to die, tied to the vast and relentless Wheel of Time.

The cyclic view of time flourished in the Greco-Roman world and formed the basis of the Stoic philosophy and its message of fortitude in defiance of fate. The Mayas most of all were obsessed by the carousel of time; they believed their fantastic calendric computations ensured the periodic return of the time-carrying gods, and ritualistic and computational errors would terminate time by putting an end to their whirligig universe.

"Time, like an ever-rolling stream, bears all its sons away," says the hymn. We think of time as a river, carrying us forward, moving from the past to the future. In the *Principia* Isaac Newton wrote, "time, of itself, and from its own nature, flows equably without relation to anything external." Time flows, said Newton, and we tend to agree. Hooke, ever at odds with Newton, was not so sure, and wanted to know where time is and how we apprehend it.

Augustine searched his soul, Newton got down to business, yet both said much the same: not the Wheel of Time but the River of Time. Both regarded time as similar to space, as a one-dimensional extension of the external world, through which we move from the past toward the future. The "now" with its memory of the past, its vivid awareness of the present, and its anticipation of the future moves through time like a bead sliding smoothly on a wire. "When we evoke time," said Henri Bergson, "it is space that answers the call." This view of spacelike time, now common in Western society, entails the notion of motion in time.

* * *

Space in the Newtonian universe existed in its own right. Distances were relative (to the positions of bodies) but space itself was absolute. The Cartesians, unlike the Newtonians, followed Aristotle and thought that space could not exist by itself. To them, space was no more than a property of matter, and where there was no matter, there could be no space. The debate between Aristotelian clothed space and Newtonian unclothed space continued into the eighteenth century and finally Newtonian ideas triumphed. The notion of time flowing "equably without relation to anything external" caused surprisingly little controversy. We inherit that lack of critical concern, and while rejecting the futility of the Wheel of Time, we are blind to the fatuity of the River of Time.

The Newtonian mechanistic universe dominated the eighteenth and nineteenth centuries. Its inhabitants shared the same public space and public time, and they all agreed on their measured intervals of space between places and measured intervals of time between happenings. This is the commonsense world we live in that accommodates the furniture of the biological and social sciences. Twentieth-century physics, however, wrenches the mind by rejecting certain aspects of it.

A meter stick that I hold in my hand, visible to all, is a measured interval of space. I cannot hold in my hand in the same way an interval of time, yet I can easily demonstrate intervals of time by asking you to listen to the ticks of a clock or watch the swings of a pendulum. Time we say is continuous and its intervals are measurable; hence time is spacelike. We talk of so many centimeters, meters, or kilometers from one place to another, of so many seconds, minutes, or hours from one happening to another, and our lives are regulated in the quantifiable domains of public space and public time.

Intervals of time may be combined with intervals of space. We combine them repeatedly when speaking of the speed of bodies. Thus six kilometers an hour – a good walking pace – is about two meters a second; 60 miles an hour is about 30 yards a second.

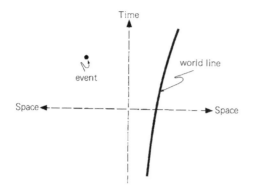

A spacetime diagram showing an event and a world line.

Scholars in the late Middle Ages at the universities of Oxford and Paris defined speed and acceleration. Speed is the distance traveled in space in an interval of time. Acceleration is the increase in speed in an interval of time. These seminal ideas of quantified motion, made clear and vivid with graphs and diagrams by medieval scholars, formed the first stepping stone to the new laws of motion.

We can easily display space and time in a diagram. Here is a vertical line representing time stretching from the past to the future, and here is a horizontal line representing space stretching from place to place. We cannot display all three dimensions of space on a blackboard or a sheet of paper, and for our purpose it is sufficient to display only one dimension. All this was more-or-less understood five centuries ago, and so far we have encountered nothing in this discussion to be alarmed about.

In the space-and-time diagram, which I shall refer to as a spacetime diagram, a point represents an event. An event is something at a place in space at an instant in time, such as the flash of a firefly. Events generally are the things we observe and are represented by points or small regions in the spacetime diagram.

At a public lecture on "Space and Time" at Cologne in 1908 Hermann Minkowski said, "Nobody has ever noticed a place except at a time, and a time except at a place." We notice events at specific places at specific times. But have you ever wondered why an individual can observe an object at the same place in space at two different instants of time, but cannot at the same instant observe it at two different places in space?

The birth of a child is an event. The child grows, experiences many events, then dies, and death is the last event. These events from birth to death when strung together form a line in the spacetime diagram. This life line, called a world line, shows the position in space of the person at each moment of time. All things that endure, such as atoms, bacteria, human beings, and stars, are represented by world lines in the spacetime diagram. Objects at rest relative to one another have parallel world lines; objects in relative motion have world lines inclined to one another. Again, this is obvious, and there is nothing to be alarmed about.

What the Newtonians said, and everyone agreed, was that between any two events the measured interval of space and the measured interval of time are the same for everybody. If one person said the separation between two events was so many meters in space and so many seconds in time, all other persons would agree and obtain the same results, even hypothetical creatures moving at very high speed in spaceships. This seemingly logical outlook changed in the first decade of the twentieth century because of the special theory of relativity.

*   *   *

The electromagnetic theory developed by James Clerk Maxwell in the 1860s was very puzzling. This elegant and powerful theory, which unified electricity and magnetism, proved that waves of light travel at a fixed speed through empty space. Light has a speed of 300,000 kilometers a second, and we now know that all electromagnetic radiation, from radio waves to X-rays, travels at this speed.

But empty space is a sort of nothing and just a vacuum. How can waves of light move at a fixed speed relative to nothing? In the nineteenth century, it seemed as if perhaps Descartes was right and

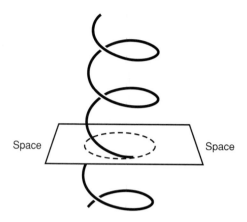

An illustration of a world line of a particle in circular motion in space. The world line is a helix, and as time advances the particle describes a circle in space.

space could not exist unless clothed in a material medium relative to which light had a definite speed. Efforts were made to conjure up a light-transmitting medium consisting of an undulating ether. It became the fashion to speak of light and other forms of electromagnetic radiation as ether waves.

If light moves at constant speed in the ether and if the Earth moves through the ether while revolving around the Sun, then by measuring the speed of light, it should be possible to detect the Earth's motion through the ether. Using the utmost precision, Albert Michelson and Edward Morley in 1887 attempted to detect the motion of the Earth by measuring the speed of light. To their surprise they found that the Earth's motion could not be detected, and the speed of light measured on Earth is the same in all directions at all times of the year.

Consider the light from a distant star. Suppose at first the Earth moves away from the star; six months later, after swinging around the Sun, the Earth moves in the opposite direction toward the star. Yet on both occasions, when the Earth has motion either away from or toward the star, the measured speed of the light from the star remains the same. The implication is that the speed of light from all sources is constant for all observers, no matter how fast sources and observers move relative to one another.

In the declining years of the nineteenth century, George FitzGerald of Ireland and Hendrik Lorentz of Holland tried to get around the problem by supposing that intervals of distance and time were altered in a way that maintained the observed constancy of the speed of light. Every thing moving through the ether had its size and periods of time altered so that light had constant speed. According to this idea, the laws of nature conspired for unknown reasons to create the impression that light, moving through the ether, had constant speed for all observers. This makeshift theory, though not very elegant, was helpful and suggestive to Albert Einstein.

Maxwell's electromagnetic theory and the puzzling constancy of the speed of light delivered the deathblow to the commonsense view of space and time. In 1905, Einstein threw away the ether and advanced the theory of relativity that has revolutionized our understanding of space and time. Instead of trying to explain Einstein's algebraic treatment, I shall use a simpler approach that offers greater conceptual insight.

*   *   *

In the late nineteenth century, the fourth dimension was all the rage. Ghosts, some writers said, were visitors from other three-dimensional spaces in a world of four dimensions. Charles Hinton, who emigrated to Princeton from Oxford, proposed in a series of essays and in his book *What Is the Fourth Dimension?* (1887) the bold idea that time is the fourth dimension. He showed how a particle in circular motion, such as a planet encircling the Sun, possesses a helical world line in four-dimensional spacetime. As the present moment, the *now* (Hinton's "plane of consciousness") advances in time, the helix describes a circular path in three-dimensional space. Hinton wrote, "We can imagine a plane world in which all the variety of motion is the phenomenon of structures consisting of filamentary atoms [world lines] traversed by a plane of consciousness." The familiar transience of things moving and changing in the observed world is explained by consciousness (an undefined metaphysical thing) moving up an observer's world line in a spacetime of fixed world lines. Hermann Weyl in 1921 echoed Hinton with the words: "The objective world simply is, it does not *happen*. Only to the gaze of my consciousness, crawling upward along the lifeline [world line] of my body, does a section of the world come to life as a fleeting image in space that continually changes." To this day we are the bewildered heirs of this metaphysics that physics accepts but cannot explain.

The world lines in the spacetime diagram, explained Hinton, are not just a convenient way of illustrating motion, but are representations of actual objects in a physical world of unified space and time. As we move along our world lines we see revealed a changing three-dimensional world of space. We travel along our world lines in a four-dimensional world like trains, and we see an ever-changing countryside of three-dimensional space. The River of Time has become a flow of consciousness.

H. G. Wells, inspired by Hinton, wrote *The Time Machine* a few years later. He explained to a different audience the idea of time as a fourth dimension. "For instance," wrote Wells, "here is a portrait of a man at eight years old, another at fifteen, another at seventeen, another at twenty-three, and so on. All these are evidently sections, as it were, Three-Dimensional representations of his Four-Dimensional being, which is a fixed and unalterable thing."

Hinton repeatedly emphasized that everything in spacetime is on display, as it was, is, and will be, and nothing changes. The cost of spatializing time as a fourth dimension is that spacetime itself is timeless and events are tenseless. Things appear to change because consciousness moves upward on sentient world lines in four-dimensional spacetime. Hinton's idea works because, although world lines are motionless, metaphysical (i.e., nonphysical) motion is permitted. We must note that the multistranded world lines of living creatures represent among other things the biochemistry of their thoughts, memories, and emotions, and in this sense thoughts, memories, and emotions are imprinted in spacetime and must be considered physical and not metaphysical.

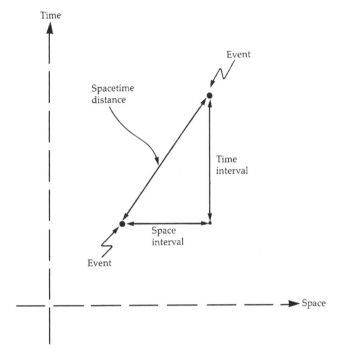

The spacetime distance between two events.

All around us is a world in action. From moment to moment and year to year things change. We occupy a world that is in a perpetual state of Heraclitean flux. Yet in the world of spacetime there is no action, nothing changes, and all is at peace and rest in a Parmenidean stillness. In our private worlds of space and time everything changes; in the public world of spacetime nothing changes. Things are displayed in spacetime in the form of events, world lines, and light rays. All that exists in the physical world is there, unhidden and on show, depicted in an unchanging and tenseless manner. All that has been, that is, and that will be *is*.

Spacetime is like a crystal ball. Every little detail of the universe throughout spacetime is on display to the percipient fortune-teller: "My dear, you have had an unhappy childhood, you are worried about your job, but do not worry, you will soon receive a letter, and will then go on a long journey, meet a tall, dark man, have two children, and live happily in another country, and die in old age." It's all there, in spacetime.

*   *   *

Einstein advanced the theory of relativity, and Minkowski, his former teacher, explained in 1908 what the theory meant in terms of the spacetime diagram. Minkowski said: "The views of space and time which I shall lay before you have sprung from the soil of experimental physics, and therein lies their strength. They are radical. Henceforth space by itself, and time by itself, are doomed to fade away into mere shadows, and only a kind of union of the two will preserve an independent reality."

Scientists had regarded the four-dimensional world as little more than a convenient graphical way of representing motion in space and time. Minkowski showed that space and time are actually fused together into a world of spacetime, and the intrinsic structure of spacetime accounts for the constancy of the speed of light for all observers.

The theory of relativity replaced the public space and the public time of the Newtonian universe with a public spacetime. Intervals of space and intervals of time between events are no longer the same for everybody; instead, intervals of spacetime and the speed of light are the things on which we now all agree. (The old invariant intervals of space and time are replaced with two new invariants: the speed of light and the spacetime interval.)

In the physical universe of today we all agree – no matter how fast we move relative to one another – on the measured values of spacetime distances between events and on the measured value of the speed of light. We do not agree on our measurements of space and time intervals, only on their combinations that give the spacetime interval. This amazing change in outlook has been forced on us by the discovery that space and time are not independent of each other. You and I share the same spacetime, but my space and your space, and my time and your time are the same only when we are at rest relative to

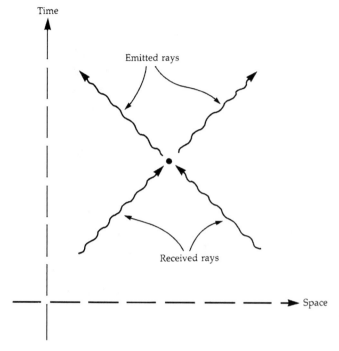

Light rays emitted and received at an event. Each light ray travels in 1 second a distance of 1 light-second in all spacetime frames.

each other. Spacetime is the new public domain, and within it we have our own individual worlds of space and time. Einstein and Leopold Infeld in *The Evolution of Physics* wrote: "The relativity theory arose from necessity, from serious and deep contradictions in the old theory from which there seemed no escape. The strength of the new theory lies in the consistency and simplicity with which it solves all these difficulties, using only a few assumptions."

*   *   *

Light in 1 second travels a distance of 300,000 kilometers, very roughly the distance to the Moon, and this distance is called 1 light-second. The distance to the Sun is 500 light-seconds. Light from the Sun takes 500 seconds to reach us, and we see the Sun as it was 500 seconds ago. Light travel time is a convenient

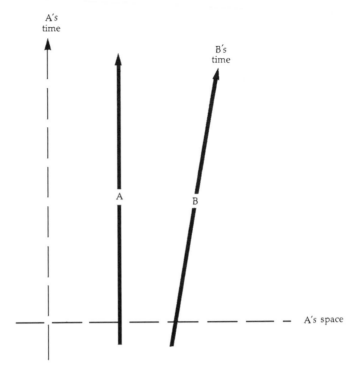

In the spacetime of special relativity, time is measured along a world line. Space is measured perpendicular to the world line, although this cannot be shown for all world lines in a diagram such as this.

way of measuring large distances, and has the advantage that we know how far we look back into the past when observing distant bodies. Nearby stars are at distances of roughly 10 light-years, or 100 trillion kilometers, and we see them as they were 10 years ago; nearby galaxies are at distances of roughly 10 million light-years, or 100 billion billion kilometers, and we see them as they were 10 million years ago.

Let us measure intervals of space in light-seconds and intervals of time in seconds. (We could also use light-years as intervals of space and years as intervals of time.) In the spacetime diagram, at each point or event, we can show light rays coming in from the past and light rays going out into the future. In 1 second the rays travel a distance of 1 light-second; hence they are inclined at angles of 45 degrees, as shown in the figure. The rays are always inclined at 45 degrees, because the speed of light is the same everywhere for everybody.

Each world line, according to the theory of relativity, decomposes spacetime into its own space and its own time. The time intervals are measured along the world line by a watch, an atomic clock, or by just counting heartbeats. Your time is measured along your world line by your clock, and my time is measured along my world line by my clock. The time that elapses between birth and death is the length of a person's world line. Observers, when in relative motion, do not have parallel world lines and do not share the same time.

The space that belongs to a world line is always perpendicular to that world line. World lines that are not parallel do not share the same space. On a sheet of paper we can only show space perpendicular to the vertical world lines; inclined world lines also have their spaces perpendicular, but unfortunately this cannot be shown in the same diagram because of the strange geometry of spacetime that I shall come to shortly.

If you rush past me, dashing in through one door and out through the other, your space and your time are not quite the same as mine. Your time contains part of my time and some of my space, and your space contains part of my space and some of my time. And vice versa, my time contains part of your time and some of your space, and my space contains part of your space and some of your time. We share the same spacetime, but not the same space and time, and our world lines determine our different worlds of space and time.

*   *   *

In *The Mathematical Theory of Relativity* Einstein drew attention in 1911 to a remarkable aspect of special relativity:

> If we place a living organism in a box . . . we could arrange that the organism, after an arbitrarily lengthy flight, be returned to its original spot in a scarcely altered condition, while corresponding

organisms which had remained in their original positions had long since given way to new generations. For the moving organism the lengthy time of the journey was a mere instant, provided the motion took place with approximately the speed of light.

This sums up the clock paradox, otherwise known as the twin paradox. One twin stays at home on Earth, and the other goes off on a long journey in a spaceship that travels close to the speed of light. Let us call the stay-at-home twin A, short for Albert, and the gadabout twin B, short for Bertha. After many years, B returns from her travels, and it is immediately apparent that she is much younger than her twin brother A who has stayed at home. Age in both cases is measured in the same way: by the clocks they carry, their number of gray hairs, and the number of heartbeats since birth.

Albert feels cheated: "It's not fair! We were born together, and now look at you – you are years younger." To which Bertha replies, "But we are not the same age. You are older, you have slept more times, ate more meals, and read more books." Their age difference is real, and the result of the geometry of spacetime. It is important to realize that in spacetime a straight world line is not the shortest distance between two events. Many of the surprising results of relativity spring from this fact alone.

Most of us are familiar with the Pythagorean theorem: the square of the hypotenuse of a right-angled triangle is equal to the sum of the squares of its two sides. There is a triangle in spacetime. Common sense insists that the hypotenuse of this triangle must be longer than either of its two sides. But common sense deceives us, because the Pythagorean theorem does not apply to spacetime. The geometry of spacetime is not the same as that of ordinary space. Triangles drawn on a sheet of paper deceive us concerning the properties of spacetime.

The spacetime distance between any two points is equal to the time interval squared minus the space interval squared. This minus sign rather than a plus sign was Minkowski's great contribution. When

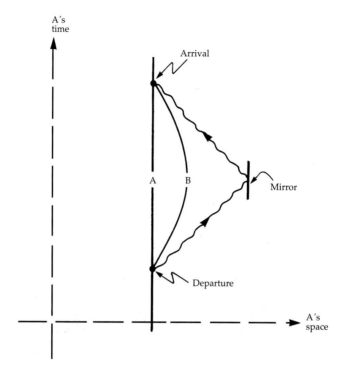

Twin A stays at home and twin B departs on a long journey at high speed. On B's returns it is apparent that B is younger than A. In spacetime B travels a shorter distance than A.

we observe two events: always your time interval squared minus your space interval squared equals my time interval squared minus my space interval squared.

We can now answer an interesting question: in spacetime, what is the distance traveled by a light ray? For example, a ray of light is emitted by one atom and absorbed by another atom. We are asked to find the distance in spacetime between the two atoms. Suppose that the ray travels for 1 second between the emission by the first atom and the absorption by the second atom. The distance in time is 1 second, and the distance in space is 1 light-second. If we square each, and then subtract one from the other, we get the result: $1 - 1 = 0$. The spacetime distance is thus zero. If the atoms are separated by

x seconds in time and hence x light-seconds in space, we get the same answer: zero distance in spacetime.

Spacetime is constructed in such a way that the spacetime distance traveled by light rays is always zero. Light rays from distant stars hurry at great speed for long periods of time across wide gulfs of space and yet travel no distance at all in spacetime. In the world of spacetime we are in direct contact with the stars.

Albert stays at home and his world line is more-or-less straight, as shown in the figure. Bertha leaves on her long journey and eventually returns and her world line is not straight but curved – it goes out and then comes back again. Suppose at the moment of her departure a pulse of light is emitted from the Earth. Suppose also that this pulse of light is reflected by a distant mirror and is received back on Earth at the moment of her return. The distance traveled by the pulse of light, measured from the Earth to the mirror and back, is zero in spacetime. Clearly, the length of Bertha's curved world is somewhere between the length of Albert's straight world line and the length of the path of the reflected pulse of light. But the reflected pulse travels zero distance in spacetime. Therefore, Bertha's curved world line is shorter than Albert's straight world line. In spacetime curved world lines are always shorter than straight world lines: they are actually shorter even though when drawn on a sheet of paper they look longer.

Time is measured along a world line. The amount of time that elapses between two events on a world line is the length of the world line between these events. Bertha goes off on her travels, and her curved world line is shorter than Albert's straight world line. The faster she travels, the more her world line approaches that of the light ray, and the shorter the time she spends on the journey. Her age when she departs is the same as Albert's and then is less than Albert's when she returns. If she could travel as fast as a ray of light, her journey would take no time as measured by her clock. In one heartbeat she could traverse the universe.

The surprising properties of spacetime are the result of the way space and time are fused together. When we combine the squared intervals of time and space we must subtract one from the other and not add them. As a result, light rays have zero length in spacetime, and curved world lines are shorter than straight world lines. With these simple facts in mind, the algebra of special relativity, formulated by Einstein, becomes smooth sailing. You, too, can be a relativist with only high-school algebra.

The effects of relativity are not apparent in ordinary life because we never travel very fast compared with the speed of light. In the laboratory, however, the effects are observed constantly. For example, an unstable particle known as the muon decays into an electron and a neutrino in about one-millionth of a second. The muon's short lifetime is measured along its world line. Suppose the muon travels close to the speed of light. In the Newtonian universe we would expect it to decay after traveling on the average 300 meters, a distance of one-millionth of a light-second. But in the modern universe it travels much farther. At a speed 99.5 percent that of light the muon travels on the average 3000 meters before decaying.

Imagine a spaceship capable of accelerating and decelerating at 1 g (equal to the acceleration caused by gravity at Earth's surface). Space travelers inside the spaceship feel a force equal to their weight on Earth owing to the acceleration. By accelerating and decelerating the spaceship, they can land on a planet in the Andromeda galaxy in only 30 years of their time. Andromeda is the large, nearby galaxy at a distance of 2 million light-years. For the travelers the total journey there and back lasts 60 years in their time. If they start when young they can return in old age. On their return, however, they will find that the Earth has aged by 4 million years.

*   *   *

Why does time flow from the past to the future? What determines the direction of the River of Time? This is the arrow-of-time riddle.

Our spacetime diagrams fail to tell us which is the future and which the past. We could turn the diagrams upside down, and they would still be much the same. A spacetime diagram, with the top labeled future and the bottom labeled past, contains little or nothing to prevent us from turning the diagram upside down so that the future becomes the past and the past the future. Yet in ordinary life the past and future are very different and there is no possibility of confusing the two. Rivers do not flow from seas to mountains and life does not begin in the grave and end in the cradle.

Most laws of physics cannot distinguish the past from the future and lack what Arthur Eddington called the "arrow of time." The arrow-of-time puzzle is similar to the problem of deciding which way up to hang a picture on the wall. An examination of the brush marks and dabs of paint fails to reveal which is the top and which is the bottom of the picture. We must stand back and look at the shapes of larger regions and even the whole picture. Much the same applies to the physical world.

On the microscopic scale there is very little in the physical world that determines the arrow of time. Two particles rush together, collide, and then rush away. When the arrow of time is reversed, they again rush together, collide, and rush away. What happens one way can happen the other. But on a much larger scale – that of plants and flowerpots – things are very different. A hot cup of coffee grows cold; an ice cube from the refrigerator melts; a drop of ink in a glass of water diffuses and disappears. Plants grow from seeds, heat travels from hot to cooler regions, energy cascades into more-dispersed and less-accessible states, organization gives way to disorganization, and the arrow of time is unmistakable.

This familiar behavior of the ordinary world, which fixes the arrow of time, is governed to a large extent by the three laws of thermodynamics. These laws have been summarized by an unknown author as follows.

(1) "You cannot win" (cannot get something for nothing because energy is conserved).

(2) "You cannot break even" (cannot keep on using the same bit of energy, because it cascades into less-useful states, and entropy increases).

(3) "You cannot get out of the game" (cannot escape because temperature has an absolute zero that is unattainable).

With these laws and their large-scale effects we can usually determine which is the past and which the future in the spacetime diagram.

But the arrow of time is still not fully understood and perplexing issues remain. For example, light comes to us from the past but not from the future, yet nothing in Maxwell's equations forbids us seeing the future. Thomas Gold has suggested that the direction of time for each world line is ultimately determined by the universe as a whole. According to this idea, the shapes and figures are not sufficient to show how to hang the picture on the wall, we must also be guided by the picture frame.

*   *   *

Spacetime is the world of fate in which every detail of life is on display and unalterably fixed. Each thread in the fabric of spacetime follows its appointed path. There are no surprises, no secrets, for all is disclosed and on full display. One's death is not hidden from sight but is depicted as plain as one's birth. Maurice Hare in *Limerick* wrote,

> There once was a man who said "Damn!
> It is borne in upon me I am
> An engine that moves,
> In predestinate grooves,
> I'm not even a bus, I'm a tram."

We might add, and not even a tram, only a tram line.

Fate is the enemy of free will. Believers in fate hold that all is unalterably ordained. Whatever I do, whether I plan or do not plan, strive or do not strive, serves no purpose and accomplishes nothing more that what already is ordained. I can alter nothing, and as John Milton said, "what I will is Fate." Human beings have limited vision, thus accounting for their illusion of free will; only the gods can peer into the crystal ball of timeless spacetime.

Meddling with fate and altering the cosmic design offends the gods of the mythic universe. The original sin of eating forbidden fruit was an act in defiance of fate. It was the first act of human free will and therefore the worst sin, for it transgressed against the divine blueprint.

The old debate of determinism versus freedom of will continues to this day with undiminished vigor. Determinism is fate in another form. The doctrine of determinism declares that everything has its cause and nothing is arbitrary. What has happened determines what happens and will happen and must have its reason and needs explanation. We seek the necessary reason and discover the sufficient explanation. Even in quantum mechanics the wavefunction evolves in a fully deterministic manner. In principle, everything is rational and nothing inexplicable. Even human behavior is deterministic. A world of predetermined events allows no room for pure chance and free choice. Imagine a world in which nothing had predestinate grooves and nothing could be predicted. It would be irrational; no person could live in it, and no society has ever devised such a universe.

Things are determined because they are the consequence of causes and the cause of further consequences. We may not always know the consequences and causes, but we believe nonetheless that they exist. We cannot predict the weather next year on a certain day, but we are confident that whatever happens on that day will have its causes. Determinism springs from the deep-rooted belief that the universe is rational and governed by intelligible laws.

Spacetime offers a godlike view of the universe: it displays the past, present, and future. This alone does not mean the universe is rational. The universe could be irrational, with everything happening randomly, ungoverned by laws, and yet still be displayed as it was, is, and will be. To make the spacetime picture rational and deterministic we must show how its events interrelate.

Freedom of will is the conviction that we as individuals have some control over our lives and are not the sport of fate and the victims of inflexible laws. All scientific, philosophical, and theological theories that explain how things work are in conflict with our personal awareness of free will. As Dr. Johnson said, "All theory is against the freedom of the will; all experience for it."

Belief in free will flies in the face of rational thought and is the essence of the Pelagian heresy (discussed in Chapter 14, The Design of the Universe). Saint Augustine in the late fourth and early fifth centuries was the architect of a form of deterministic theism. Nothing acted freely and everything necessarily obeyed the grand design. Freedom of human will contradicted the omnipotent will of God, and human beings followed their predestinate cosmic grooves as ordained by God. "Give me what thou commandest, and command what thou wilt," said Augustine in the *Confessions*. The cost of a rational, explicable universe – any rational, explicable universe – is the loss of free will. Freedom of will must be condemned as an illusion of the deceived senses. But Pelagius, a British monk of the same period, preferred a universe in which not the will of God but the will of human beings accounted for human behavior and sin. Pelagius rejected original sin and was condemned as a heretic. All persons who claim freedom of will and deny the absolute determinism of the universe in which they live are guilty of the Pelagian heresy. I am myself a Pelagian heretic.

10　What Then is Time?

Everything is spread out in time. Things stretch away into the recent past as recalled in our memories and newspapers, and into the distant past as recounted by historians and geologists. They also stretch away into the near future as anticipated in our plans and foretold by fortune tellers, and into the distant future as predicted by geologists and astronomers.

Say no more of time! If you want a peaceful mind go no farther. Every step in quest of understanding time leads to greater bewilderment. Much of the problem is that our languages inadequately express our experiences of time.

"What, then, is time?" asked Augustine of Hippo in the *Confessions*. "If no one asks me, I know what it is. If I wish to explain what it is to him who asks me, I do not know." He viewed time as a continuous temporal sequence from the past to the future, from Creation in the beginning to Judgment in the end. Time thus displays a historiography ordained by either God, fate, or natural law. This is much the same as our present commonsense general view. It caused him much perplexity, some of which is expressed by Austin Dobson in *The Paradox of Time*:

> Time goes, you say? Ah no!
> Alas, Time stays, we go.

Some of the problem is easily stated: Nothing displayed in time can change! If you think of time in terms of space, as an extension, a sort of one-dimensional space, with everything displayed in it, such as birthdays, anniversaries, holidays, then everything has its fixed moment in time and cannot possibly change. My birthday occurred on a certain day in January in a certain year, and nothing can ever change that date. By spatializing time, time becomes timeless, and nothing in it changes. We see this clearly in a spacetime diagram. Things are fixed in the form of world lines, events, and light rays. How can things change when fixed in time? The solution at first seems obvious: Time "flows equably" said Newton. Our sense perceptions are limited to a short period called the *now*, or the present moment, and the *now* moves in time traveling from the past to the future. Thus the observed world, like the countryside seen from a railway carriage window, constantly changes and creates an awareness of change.

The notion of movement in time, if we think about it, makes absolutely no sense. It compels us to ask: at what speed do we move in time? But already time has been used once, and now it must be used again to tell us our speed in time: so many seconds a second! The idea seems absurd, yet we use it constantly, and even think of moving in time at different speeds, as in the poem by Guy Pentreath:

> For when I was a babe and wept and slept, Time crept;
> When I was a boy and laughed and talked, Time walked;
> Then when the years saw me a man, Time ran,
> But as I older grew, Time flew.

If the *now* moves in time, there must be a second time, said John Dunne in 1927 in *An Experiment with Time*. But in this second time we have motion also, and hence there exists a third time, implying yet a fourth, and then a fifth, and so on, without limit. Most persons find Dunne's notion of serial time unappealing. The philosopher C. D. Broad in 1935 dismissed serial time on the grounds "We can hardly expect to reduce changes of Time to changes in Time," and Gerald Whitrow in *The Natural Philosophy of Time* (1980) makes the remark, "time is not itself a process in time."

Nothing physical changes in spacetime. World lines, events, and light rays are all displayed in a fixed state. The movement of the *now* cannot be a physical movement.

We see a world of change, said Charles Hinton in the nineteenth century in *What is the Fourth Dimension?*, because consciousness moves along a world line thereby disclosing scenery that seems to change. Consciousness is a metaphysical thing and its movement in time must be regarded as metaphysical and not physical. The *now* is a traveling metaphysical time machine. Hermann Weyl in *Space–Time–Matter* early in the twentieth century said much the same: "Only to the gaze of my consciousness, crawling upward along the lifeline [world line] of my body, does a section of the world come to life as a fleeting image...." James Jeans wrote in 1936 in *Scientific Progress*, "The tapestry of spacetime is already woven throughout its full extent, both in space and time, so that the whole picture exists, although we only become conscious of it bit by bit."

The spatializing of time simplifies its properties and in return we obtain in history an orderly uniform progression of events and in science a mathematical framework of great power and elegance. The problem of the *now* moving in time, creating the illusion of a world in transience, is transferred from physics to metaphysics. Consciousness and the speed at which the *now* of consciousness moves in spacetime become metaphysical subjects that scientists gladly leave to philosophers.

We live in the *now*. We have a vivid awareness of things in the present, and have memories of things in the past and anticipations of things in the future. The *now* in which we live with its awareness of the past, present, and future has become in language and science a segment of time that paradoxically moves in time, and inexplicably moves unidirectionally. When we think of ourselves in time (or as world lines in spacetime), the *now* vanishes, and when we try to put it back with its awareness of things changing we think of it as a segment of time moving in time (or along a world line) like a spark along a fuse. It's absurd, it's preposterous, and most of us avoid thinking about it, except for the Eleatics among us who add to the absurdity by telling us that time is an illusion.

*   *   *

Greek philosophers in the sixth and fifth centuries B.C. identified two aspects of time – *being* (the extension aspect) and *becoming* (the transience aspect) – that to this day remain unreconciled. Let me try and explain.

Heraclitus of Ephesus said everything changes, nothing endures, and the world consists of an endless flux of things always changing, always becoming, and we never step into the same river twice. He emphasized the transience aspect of time. At about the same time Parmenides of Elea said nothing changes, everything endures, and the world consists of an "invariant sphere of being" in which the past, present, and future are simultaneously displayed. He emphasized the extension aspect of time. Heraclitus said the world consists of the transient acts of things becoming and an unchanging state of being is an illusion of the mind. Parmenides said the world consists of an unchanging state of being and the transient acts of things becoming are illusions of the senses.

In the sensible world of everyday life, we experience transient happenings and common sense sides with Heraclitus. To say that nothing actually changes contradicts experience. But in language and science we are Eleatics and suppose that things exist in time, in the past, present, and future, and explain transience by invoking a mysterious motion in time or motion of time itself.

In language and science we spatialize time as a sort of Parmenidean state of being, and lapse into perplexity when we suppose we have exhausted the basic properties of time. By spatializing time we stress its innate extension aspect but ignore its innate transience aspect. The impossibility of expressing transience in spatialized (extended) form tempts us in to thinking that transience is not a fundamental property of the external world but an illusion that is psychologically or metaphysically peculiar to the observer. In response to this popular Eleatic belief, Gerald Whitrow asks, "How do we get the illusion of time's transience without presupposing transient time as its origin?" Transience is an irreducible property of time; and when dismissed in one form, it always reappears in another.

Science has that delightfully simple answer: the transience is not physical but metaphysical and is therefore not a scientific problem. But suppose that science is wrong. What then? Suppose that Bertrand Russell was also wrong when he wrote in *Mysticism and Logic*: "time is an unimportant and superficial characteristic of reality. Past and future must be acknowledged to be as real as the present, and a certain emancipation from slavery to time is essential to philosophical thought." If science is wrong then so are many other branches of knowledge, and one might go further and say that even our languages are misconstructed. Since the time of Plato most thinkers in the West have accepted the principle that the mind explores an abstract world of timeless reality and our deceived brains or minds account for the transient character of personal experience.

I am inclined to suggest that in our search for physical reality we should question the basic assumptions of this Eleatic philosophy.

*　　*　　*

The extension and transition aspects of time are equally important. In one sense, we are aware of time as a state of being, an extension, throughout which events are distributed. This is the aspect of time that has been spatialized and woven into the modern fabric of spacetime. In another sense, we are also aware of time as an act of becoming, of one state of being transforming into another state of being, of one vista of past, present, and future dissolving and reforming into another vista of past, present, and future. It seems that the tapestry of being is rewoven in each act of becoming. This aspect of time defies a purely spatial representation. We omit transience from science because we have not learned how to express it in a fundamental physical form. To dismiss the act of becoming as an illusion throws away a vital aspect of time and greatly oversimplifies the world in which we live. We pretend there is no problem by banishing consciousness and transience to the disneyland of metaphysics.

In the River of Time we have a clear state of being that consists of the past, present, and future. This is the aspect of time that is spacelike. The future supposedly stretches away ordained and as clear and detailed as the past. But of course in reality the future never is clear.

The act of becoming defies description in spatial terms, and by denying that it exists we create as a result the perplexing motion-in-time problem and the notorious arrow-of-time riddle. The *now* is reduced to a mere segment of time endowed with metaphysical motion. But metaphysical motion still encounters the question: at what speed does consciousness move in time? And why only in one direction? Spatialized time by itself fails to express our awareness of the transience of things becoming. When we try to express the act of becoming, it melts away and we are left with only a paradoxical movement in time, and with time that supposedly flows, which cannot flow.

By failing to put the *now* complete with its transient act of becoming into the physical world picture, we fail also to put our consciousness into the physical description, even in its simplest representative form. We fail to put ourselves into the physical universe as experiencing individuals. A physical universe that experiences nothing is one in which there is no transience. Possibly the next major step in the grand design of universes will be the discovery of a more sophisticated way of representing time and consciousness.

Everything is laid out in spacetime. This is the being aspect of time. The whole of spacetime is the *now*. The spacetime state of being dissolves in an act of becoming and a new spacetime state of being emerges. Each state of being is a *now*, and each *now* contains its past, present, and future.

We remember the past and foresee the future in each *now*. The *now* does not move from the past to the future. Each *now* is a state of being that contains the past, present, and future. Each *now* collapses and is replaced by a new *now* that contains again the past, present, and future. Alternatively, one might say a state of being collapses in an act of becoming and is replaced by a new state of being. Each state of being displays the present emerging from the past and evolving like a super wavefunction into many potential futures. In an act of becoming a state of being decays and collapses and forms a new state of being in which some of what was previously potential becomes actual. Each *now*, let us face it, is a state of conscious awareness.

*   *   *

In the tenth and eleventh centuries, Arab atomist philosophers of the Kalam – known as the Mutakallimun – rejected the Aristotelianism of more orthodox Muslim theology. The world, they said, has no natural laws. It consists of noninteracting atoms governed supernaturally by the will of the sole agent (the supreme being). The Jewish scholar Moses Maimonides, in *The Guide for the Perplexed*, discussed critically the speculations of the Mutakallimun and this twelfth century work serves as a primary source of information on the Kalam philosophy.

Bakillani of Basra, who lived in Baghdad where he died in 1013, hit on the idea that time also is atomic, and in each atom of time the sole agent annihilates the world and recreates it in slightly different form. On the subject of time atoms, Maimonides wrote, "An hour is, e.g., divided into sixty minutes, the minute into sixty seconds, the second into sixty parts, and so on; at last after ten or more successive divisions by sixty, time-elements are obtained that are not subjected to further division, and in fact are indivisible."

The Mutakallimun sought to demonstrate the slavish dependence of the world on the will of the sole agent. The world itself had no intrinsic power of self-explanation, everything depended on divine will. Bakillani's acts of creation, which continually reform the external world, posed a cosmological problem: If the countless creations, each isolated in its atom of time, lack connection, how do human beings succeed in arranging them into an orderly sequence? The Kalam solution anticipated the theory of occasionalism proposed in response to the Cartesian mind–body problem: the sole agent creates in each atom of time two parallel worlds, a material world and a corresponding mental world.

It is interesting that the Kalam theory partly reconciles the dual aspects of time. In each time atom the world stretches away in space, and in addition has a past and future stretching away in extended time. This corresponds to a state of being in which nothing changes. The world of space and time then dissolves and a reconstructed world of space and time emerges. Again, in the new atom of time nothing changes. Change consists of successive transformations from one time atom to another time atom, and these transformations act as the transient aspect of time.

Shorn of medieval theism and its denial of natural law the Kalam theory describes moderately well how we experience time. When I attempt to describe what happens, I find the *now* always contains the recalled past and anticipated future and both stretch away in extended time; I live always in the *now*, never in the past and never in the future; I experience no motion in time, and notice only that the past, present, and future change from one momentary state of conscious attention to another, from one *now* to another *now*, and in this way I experience transient time. On substituting *now* for time atom my experiences seem to agree with the Kalam description of time.

Being and becoming are conjugate aspects of time. The philosopher Alfred Whitehead wrote in *Process and Reality*, "In every act of becoming there is the becoming of something with temporal extension, but . . . the act itself is not extensive." This sounds very much like saying the acts of becoming (in transient time) create states of being (displaying extended time).

Conjugate time is analogous to reading a book. A page of writing is like a state of being in which everything is laid out on display. Turning the page is like an act of becoming in which one state of being dissolves and is replaced by a new state of being.

*   *   *

The theory of conjugate time, when updated with modern ideas, jolts the imagination. It requires that every act of becoming transforms all spacetime; transforms the here and there and the now and then.

The past today is different from the past yesterday if only because it now contains what happened since yesterday. A state of being extends throughout spacetime and an act of becoming transforms one state into another state. Thus in each act of attention not only remembered things of the recent past and anticipated things in the near future change but also the things of the distant past and future. What happened in 1900 was physically not exactly the same last year as this year. History is not a concrete sequence of fixed events.

I have argued that time possesses conjugate aspects: extension (being) and transience (becoming). The first can be spatialized in language and science but the second cannot. The theory of conjugate time attempts a reconciliation of these dual aspects in which each is equally important.

The *now* is not a segment of time that moves in time but a configuration of the world throughout space and time. Each configuration corresponds to an observation – a conscious act of attention – and the manifold of configurations in acts of becoming constitutes a universe of immense complexity.

In conjugate time we must distinguish between the old *now* (or present moment or "specious present") and the new *now* (time atom or "chronon"). The old *now* is a window spanning an interval of extended time that mysteriously moves from the past to the future. The new *now* is the whole world displayed throughout space and extended time in which the past stands clear and certain (or so it seems) and the future looms vague and potential of many forms.

The duration of a time atom, when measured in extended time, varies and depends on the observation. The span of the "present moment," in which past certainty decays into future uncertainty, is a plausible measure of the duration of a time atom. In ordinary human affairs the decay varies from a fraction of a second to several seconds, even minutes, and this variation may account for our experience that intervals of extended time sometimes seem comparatively long or short and we say that time passes slowly or quickly.

On the atomic scale the decay lasts typically one hundred millionth (10^{-8}) of a second. Most situations consist of numerous systems in states of decay and the duration of the specious present depends on the system and the observation and may last a billionth of a second or a billion years.

The wavelike aspect of a particle complements its corpuscular aspect. Similarly, the extension aspect of time complements the transition aspect. The collapse of the wavefunction seems not unlike the collapse of a state of being in an act of becoming. A state of being displays the past evolving into the present and fading away into a superposition of future probabilities. The state collapses, an observation is made, and a new state emerges equipped with a new past and a future that was foreshadowed in the previous state.

*　　*　　*

The duality of temporal extension (fixed events) and transience (changing events) creates confusion and the age-old paradoxes and riddles of time. In our languages and in sciences we resolve the confusion by accepting extension as real and rejecting transience as unreal. We accept extended clock and calendric time as physically real and treat transient time as a psychic addition having nothing to do with the real world. The transience aspect of time defies description with spacelike imagery and metaphor; it lacks an obvious system of measurement and for this reason, and for the sake of simplicity, it has been omitted from the mathematization and mechanization of the world.

Conjugate time introduces transience into the physical description. By so doing it reveals a world of much greater complexity than we normally imagine.

11   Nearer to the Heart's Desire

We have a picture of a seamless spacetime projecting into the space and time of each observer's world line. Though elegant and economic, in one sense it differs little from the Newtonian picture. Space in the Newtonian scheme was just a sort of nothing (a sideless box) spanning everything, and time was a similar sort of nothing in which everything also had location. In the theory of special relativity both came together to form an expanse of spacetime containing everything that again was just a sort of nothing (just a bigger sideless box).

Then in 1916 Einstein advanced the theory of general relativity and the picture changed dramatically. (How dramatically was not fully realized for many years.) Spacetime lost its state of nothingness and acquired a tangible physical reality. Gravity ceased to be a mysterious astral force acting instantaneously at a distance and became a property of dynamic curved spacetime.

In the new scheme spacetime itself guides the heavenly bodies and the old astrological action at a distance turned out to be the curvature of space and time combined into spacetime. We now have a spacetime that pulls and pushes and transmits shivers and shakes at the speed of light. We cannot eat spacetime, but it can be hit, and can hit back, and can eat us if we stray too close to a black hole. Spacetime in general relativity springs to life and becomes an active participant in the physical universe.

In the Newtonian scheme something curious and rather interesting about the nature of gravity points in the direction of general relativity. On Earth we feel the pull of gravity. In an accelerating vehicle we feel a force that seems very much like the pull of gravity. If we wish to contemplate nature undisturbed by the pull of gravity, we may follow Arthur Eddington's advice and "take a leap over a precipice" and enter into a state of "free fall." Here in a nutshell are the essential ingredients of the equivalence principle that points the way to general relativity.

Imagine a spaceship in the depths of space equipped with all kinds of scientific apparatus. The spaceship is without windows and the experimenters inside cannot see what is happening in the outside world. We suppose the spaceship at first is far from any star and undisturbed by gravity. It moves freely at constant velocity. With their various instruments the experimenters find that they are unable to determine how fast and in what direction their windowless spaceship is moving. To them it seems motionless. A ball, for instance, floats above the floor and remains stationary inside the moving spaceship. Any experiment performed inside the spaceship yields results always in conformity with special relativity theory. Gravity is absent, and space is uncurved and "flat." By space I mean spacetime, but "space" is easier to think about.

The spaceship eventually approaches a star, swings around the star in a curved orbit and moves away in a new direction. While this happens the ball continues to float above the floor and remains stationary inside the spaceship. (Complications owing to gravity variations within the spaceship need not detain us in this discussion.) The experimenters perform experiments with various instruments and remain totally unaware of the gravitational pull of the nearby passing star.

The spaceship moves freely. It is a free-falling system following a trajectory of such a nature that the force produced by its acceleration always cancels exactly the gravitational force produced by the star. This quite remarkable state of affairs lies at the heart of the Newtonian scheme of celestial dynamics. Perhaps you have difficulty believing that in free fall the force of acceleration exactly cancels gravity. You need not jump over a precipice to be convinced of its truth; you might only find that air is a very resistive medium. Astronauts while orbiting about the Earth in free fall have shown us on television screens how objects float in a weightless state inside their space vehicles. The astronauts also experience this weightless state because gravity is nonexistent in all free-falling systems.

Our experimenters out in the depths of space remain under the impression that their windowless spaceship, while passing a nearby star, continues to move at constant velocity. They still think their spaceship is still far from any star and undisturbed by gravity. Their experiments give continually the same results and they continue to use special relativity as the theory that explains their observations.

We on Earth cannot feel the pull of the Sun. The Earth, in free fall, moves around the Sun always in such a precise way that its motion cancels the Sun's gravitational pull. We feel the pull of the Earth's gravity because we, as surface-dwellers, are not in free fall about the Earth. The principle of equivalence asserts that the gravitational pull of a body is annihilated within any system that free-falls about the body. If there is a grand theory of gravity, this principle tells us that it must simplify to special relativity theory in all free-falling systems.

The Newtonian universe failed to provide a grand theory of gravity. It had the serious defect that gravity acted instantaneously everywhere. When an apple fell to the ground, all places in the universe received the news simultaneously over the gravity network. Newtonian gravity ignored the speed limit of light and, in principle, could be made to violate causality and do impossible things.

As an illustration, suppose that A (for Albert) and B (for Bertha) are in separate spaceships fleeing side by side from enemy X. In this science-fiction scenario of star wars it will not strain the credulity of the reader if we further suppose that all characters have Newtonian gravity guns that shoot at infinite speed.

Enemy X fires and destroys Albert. The act of firing and the act of destroying Albert occur simultaneously in enemy X's space. But Albert and Bertha have their own space and time, and for them these acts do not occur simultaneously. In fact, the act of firing by X occurs after the act of destroying Albert. On seeing the destruction of Albert, Bertha fires back immediately, and by destroying enemy X, she saves Albert. An effect is thus canceled after it occurs. This violates the sacred law of causality: murder once done cannot be undone. Causes cannot follow after effects.

Let us assume that a second enemy Y is stationary in the neighborhood of X. Our tale of space wars now takes a bewildering turn. Enemy Y sees the destruction of X and immediately fires back and by destroying B saves X; A thereupon destroys Y and saves B; X then destroys A and saves Y; B now destroys X and saves A; and so on repeatedly. Faster-than-light transmission enables us to perform miracles and consequently is impossible in a rational universe.

Einstein was confident of the existence of a grand theory according to which gravity travels at finite speed, a grand theory that simplifies to special relativity in systems in free fall, a general theory applying to all observers everywhere, unlike the special theory that applies only to observers in gravity-free regions.

* * *

Euclid in the third century B.C. at the Museum of Alexandria brought together the geometrical knowledge of the ancient world and established the remarkable Euclidean system of geometry. His geometrical system was not merely a Babylonian–Egyptian ragbag of rules and recipes but an analytical body of knowledge developed from a few explicitly stated assumptions. If you accepted the assumptions as both reasonable and obvious, then step by step, in a logical progression, you were compelled to acknowledge the rest.

An important basic assumption of the Euclidean system of geometry is the parallel postulate. Equidistant straight lines are parallel lines that are everywhere separated by a constant distance. The postulate states that through any point a straight line may be drawn equidistant from another straight line. In our bones we know that when two parallel straight lines extend to great distances they remain equidistant and never intersect each other. The parallel postulate cannot be shown to be true with absolute certainty because all human experience occurs within a limited region of space. It seems eminently sensible and was accepted without question by most geometers from the time of Euclid until the nineteenth century. A few geometers felt uneasy and tried to derive the parallel postulate from a more basic assumption. But all attempts failed. We now know that the parallel postulate is fundamental to Euclidean geometry. It enables us to distinguish the geometry of Euclidean space from the geometries of non-Euclidean spaces.

We find, when our observations are limited to what happens in small regions, that many non-Euclidean spaces have geometries closely resembling the geometry of Euclidean space. Small triangles and circles drawn in these spaces look much the same as our familiar triangles and circles. The simplest way to recognize such a non-Euclidean space is to observe what happens to figures covering large regions of space. This is not easy when large regions extend far beyond the range of normal experience. Can we be fully confident that two straight lines, seemingly parallel to us in our local region of space, when extended beyond the limits of large telescopes, will stay parallel in distant regions?

Curved and uncurved two-dimensional surfaces can be visualized with moderate ease. Consider first a flat surface of unlimited extent. This is Flatland that possesses Euclidean geometry. Parallel straight lines drawn by two-dimensional Flatlanders in a small region, when extended to great distances, remain equidistant and do not intersect in either direction. The parallel postulate holds true in Flatland. But equidistant straight lines drawn in curved surfaces, when greatly extended, do not remain equidistant. Large triangles and circles in Curveland are not exactly the same as similar figures in Flatland.

The uniformly curved surface of a sphere, for example, possesses non-Euclidean geometry. A small region of a spherical surface is almost flat. The smaller a region in Sphereland, the flatter seems the surface in that region. If we were two-dimensional Spherelanders living inside a very small region of Sphereland, we could easily believe its geometry is Euclidean and that Sphereland is actually Flatland. We might even find it difficult to imagine Sphereland as not being Flatland. This is analogous to the situation in our world of three-dimensional space. We live in a comparatively small region and we believe that space has Euclidean geometry. We find it difficult to imagine what curved three-dimensional space is like.

A straight line drawn on the surface of a sphere is a great circle. A Spherelander traveling in a straightforward direction follows a great circle and eventually returns to the starting point. This creature starts off in one direction and returns from the opposite direction. A great circle divides a spherical surface into two hemispheres. It is like the equator, or a line of fixed longitude on the Earth that passes through both poles. Consider a second straight line, close to the first, also of fixed longitude. In a small region at the equator the two lines appear to be parallel. When extended, however, both lines intersect at the poles and cannot therefore be parallel. All straight lines on the surface of a sphere intersect one another, and the parallel postulate fails to apply.

Curved two-dimensional space is easy. Much less easy to imagine is curved three-dimensional space. This space may have analogous spherical geometry. If we lived in such a curved space and traveled in a straight-forward direction, we would ultimately return to our starting point.

Through a point in flat space there passes one parallel, and only one, to any given straight line. This is the Euclidean postulate. Through a point in spherical space of uniform curvature there passes no parallel to any given straight line. Through a point in hyperbolic space of uniform curvature (which I have not discussed) there passes many parallels to any given straight line. The parallel postulate of Euclid uniquely distinguishes Euclidean geometry and fails to apply to other spaces not only of uniform but also nonuniform curvature.

In our small part of the physical universe we think normally in terms of flat, three-dimensional Euclidean geometry and have utmost difficulty trying to imagine curved non-Euclidean space. Immanuel Kant went so far as to declare that non-Euclidean space is inconceivable and hence impossible. Euclidean geometry, he argued, is a priori (prior to experience) and "an inevitable necessity of thought." But he was wrong. On large scales the physical universe need not conform to ideas that derive from human experience on small scales. New experiences lead to novel ideas, novel ideas to new experiences.

*   *   *

In the first half of the nineteenth century, Johann Gauss at the University of Göttingen, a trailblazer in many fields of mathematics, formulated techniques for studying curved surfaces. If we were two-dimensional creatures living in a curved surface, unaware of a third dimension, we would survey and determine the geometry of our surface with the methods developed by Gauss. But of all the mathematicians who have contributed to our knowledge of non-Euclidean geometry we remember most Gauss's brilliant young colleague, Bernhard Riemann. Riemann explored the metric properties of continuous spaces of two, three, and more dimensions and formulated the general equations defining their intrinsic properties, such as curvature and the variation of curvature. Euclidean space is unique in having zero curvature; all non-Euclidean spaces have curvature. Curvature is either uniform (the same everywhere as in Sphereland) or nonuniform (not the same everywhere as in Hillyland).

Riemann foresaw the possibility of a close relationship between geometry and physics. His studies on the curvature of space seemed at the time excessively abstract and divorced from reality. "Only the genius of Riemann, solitary and uncomprehended, had won its way by the middle of the last century to a new concept of science," said Einstein.

The mathematician William Clifford, who died when still a young man (Riemann also died when still relatively young), translated Riemann's work on geometry into English. In *The Common Sense of the Exact Sciences*, published posthumously in 1885, he championed the idea that geometry and physics are interconnected:

> We may conceive our space to have everywhere a nearly uniform curvature, but that slight variations of curvature may occur from point to point, and themselves to vary with time. These variations

of the curvature with time may produce effects which we not unnaturally attribute to physical causes independent of the geometry of our space. We might even go so far as to assign to this variation of the curvature of space "what really happens to that phenomenon which we term the motion of matter."

Clifford predicted the possibility of curvature waves (now referred to as gravity waves) and surmised, "this property of being curved or distorted is continually being passed on from one region of space to another after the manner of a wave." A germinal idea was in the air. But special relativity had yet to be discovered and the development of spacetime into a Riemannian world of dynamic curvature lay thirty years ahead.

*　　*　　*

Albert Einstein, born in Germany in 1879 (the year Clifford died), was an imaginative child whose teachers regarded him as a backward and inattentive. He read widely, developed an independent outlook, and was mainly self-taught. Newton with his "silent face" and Einstein with his retiring manner shared much in common. Both had mystical religious and metaphysical views; both were not particularly distinguished as teachers ("there is too much education altogether," said Einstein); both were pestered by distracting adulation that made further scientific work difficult ("it is unfair and in bad taste," said Einstein). Each in his way had an extraordinary intuitive grasp of physical processes, and each pondered deeply for many years before he produced his great theory of gravity.

Einstein's theory of general relativity reached final form in 1916, and to scientists and many sections of the public it seemed at last that Omar Khayyam's dream had come true:

> Ah love! could thou and I with him conspire
> To grasp this sorry scheme of things entire,
> Would not we shatter it to bits and then
> Remould it nearer to the heart's desire!

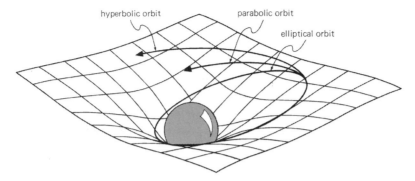

hyperbolic orbit   parabolic orbit

elliptical orbit

A stretched rubber sheet, depressed by a heavy ball, illustrates how deformed space acts as a gravitational force. (E. Harrison, *Cosmology: The Science of the Universe*, 2nd edition, Cambridge University Press, 2000.)

The Newtonian universe with its Euclidean geometry and mysterious gravity was finally shattered and remolded into a universe of varying geometrical curvature. The curved paths of the heavenly bodies moving under the influence of gravity in the Newtonian universe of flat space became the straight paths (or geodesics) in the Einstein universe of curved space.

Einstein's equation of general relativity looks harmless enough:

$$R_{ij} + \frac{1}{2} g_{ij} R = T_{ij},$$

but is actually ten equations expressed in compact notation. Terms on the left side of the equation deal with the curvature of space, and terms on the right deal mainly with matter and energy. Simple-mindedly we may think of matter as a form of stress that produces geometry strain:

geometric strain = material stress.

The strain or curvature of space is produced by the stress of matter (or energy). Local space is deformed by distant as well as local matter. Matter influences the curvature of space and the curvature of space reacts back and influences the motion of matter.

A flexible rubber sheet stretched flat illustrates what happens. The flat sheet represents uncurved space in the absence of gravity. Ball bearings rolled on the flat surface follow uncurved trajectories.

A heavy ball placed in the center of the sheet produces a depression and the rubber sheet is then everywhere curved. This represents the curvature of space produced by a star and shows how matter affects both local as well as distant curvature. The curvature near the ball is large (where gravity is strong) and far from the ball is small (where gravity is weak). The curvature diminishes with distance and at large distances the sheet is almost flat. Ball bearings rolled on the curved surface follow curved trajectories similar to those of planets and comets under the influence of the Sun.

General relativity simplifies to Newtonian theory when gravity is weak (space is almost flat) and also bodies have speeds that are small compared with the speed of light. In the Solar System, where gravity is moderately weak and the planets and comets move at comparatively low speeds, the Einstein equation reduces to the Newtonian equations of motion and Newtonian gravity. Small discrepancies remain, however, and years of research have been devoted to detecting in the Solar System these residual effects of general relativity. The results, in good agreement with predictions, inspire confidence in the validity of Einstein's theory of gravity.

General relativity is not an easy theory to use. Consider two stars in the Newtonian universe. The gravitational force that each exerts may be calculated as if the other did not exist. At any point in space the separate forces exerted by each star can be added to give the combined force exerted by both stars. But in general relativity this is no longer true. The curvatures produced independently by each star cannot be simply added to give the combined curvature produced jointly by both stars. Spacetime self-interacts in a curious manner. The curvature produced by one star alters the curvature produced by the other.

All forms of energy have mass; thus heat, which is a form of energy, has mass. A kettle of boiling water weighs a billionth of a gram more than when the water is cold. Space curvature is another form of energy. The curved space around a star has energy and therefore has its own effective mass.

The curved space (or rather curved spacetime) around a star, because of this "stress" energy and its equivalent mass, acts as a source of additional gravity. Thus curved space is itself a source of further curvature. In the Newtonian universe gravity itself is not a source of gravity, but in the Einstein universe curvature is a source of curvature. The curvature of space produced by one star, because of its associated energy and equivalent mass, interacts with and modifies the curvature produced by the other star, and the combined curvature is more than the sum of the separate curvatures of the two stars. Self-interaction of space is the essence of general relativity.

The energy of the curved space around the Sun produces more curvature and contributes to the Sun's distant gravitational pull on the planets. For this reason the Sun's gravity fails to obey precisely the inverse square law, and the planets do not follow exactly Kepler's elliptical orbits.

Consider two stars circling around each other. They move in curved orbits because of the curvature of space (or rather of spacetime). The orbiting stars cyclically strain and warp the surrounding space. This cyclic warping streams away as gravity waves at the speed of light. The waves transport energy, and the drain of energy from the binary system causes the two stars to spiral slowly toward each other. Such an effect has been observed and studied by Joseph Taylor while at the University of Massachusetts in the case of a binary pulsar system.

Einstein's equation can be viewed as a wave equation giving a wavelike description of gravity showing how curvature disturbances (gravity waves) travel at the speed of light. Emily Dickinson in *Time and Eternity* wrote,

I never saw a moor,
I never saw the sea;
Yet I know how the heather looks,
And what a wave must be.

We have never seen gravity waves, yet we know what they must be, and we have little doubt that they exist.

The ancient atomists believed that nothing exists except atoms and the void. Aristotle and later Descartes insisted that space could not exist by itself as an empty insubstantial void and must therefore be dressed in material raiment. Interestingly, space is not only dynamically curved but also suffused with energy. Quantum mechanics shows that space is densely populated with virtual states. On a submicroscopic scale space seethes with particles coming into and out of existence too briefly to produce a gravitational effect. Or perhaps not entirely. Ninety percent of the universe consists of undetected dark energy. Perhaps the virtual states of the vacuum are not entirely virtual and there is some remnant gravitational effect. Had Aristotle and Descartes known of the physical properties of space they might have viewed this as proof that nature abhors a pure vacuum. But actually it is the Newtonians who have been vindicated. They believed space was reified by spirit and thereby real in its own right. The nature of space in general relativity theory and quantum mechanics seems more etheric in the Newtonian sense than material in the Cartesian sense.

Out in the depths of space something has happened in our free-falling windowless spaceship. The experimenters have emerged through an open hatch and are now gazing at the external world of whirling bodies and orbiting systems. Everything they see in the firmament acts in accord with the grand picture of general relativity. The free-falling experimenters in their own locally flat space look out and see celestial bodies following curved paths as if under the influence of a mysterious long-range astral force called gravity. But in fact the freely falling bodies are following straight paths in curved spacetime.

*   *   *

Normally, gravity is weak as in the Solar System and the Newtonian picture suffices for most of science and much of astronomy. But gravity is sometimes strong and produces astonishing effects. Bodies "having not the law," said Saint Paul, "are a law unto themselves." Black holes – monsters of the deep – having not the Newtonian law are subject to the higher law of general relativity.

Newtonian gravity warns us that odd things happen when stars are either dense or massive. John Mitchell, rector of Thornhill in Yorkshire, pointed out in 1784 that a particle must move at least one-five hundredth the speed of light to escape from the Sun's surface. He argued that if a star had the same average density as the Sun (slightly greater than the density of water), and a diameter more than five-hundred times greater than the Sun's diameter, then not even light could escape from the surface of the star. "All light emitted from such a body would be made to return to it by its own power of gravity," he said. The star, he thought, though invisible would still be detectable because of the effect of its strong gravity on the motions of satellites and nearby stars. The astronomer William Herschel was intrigued by Mitchell's argument and thought that many luminous interstellar clouds could be interpreted as regions of light trapped by gravity.

Newtonian theory implies the possibility of the strange bodies we nowadays call black holes, and general relativity theory enables us to understand them. A black hole is born when a star collapses to extremely high density. Imagine a shrinking star. Gravity at its surface steadily grows in strength and ultimately reaches a limit. Gravity is the old name for what is now known as the curvature of space. As the star shrinks, the curvature of surrounding space increases, and ultimately space becomes sufficiently curved to actually enclose the star. It has then become a black hole from which not even light can escape. The Sun would become a black hole if its diameter of more than one million kilometers shrank to just six kilometers. The size of a black hole varies in proportion to its mass. The Earth would have to shrink to the size of a golf ball to become encapsulated in its own space. The larger a black hole, the lower its density, and very large black holes have very low densities. A black hole one billion times the mass of the Sun has a size roughly that of the Solar System and a density about the same as ordinary air.

Inside a black hole everything collapses rapidity in a crescendo of rising density. But to a person outside at a safe distance nothing happens; the black hole seems frozen in a state of suspended animation and at its surface time stands still. (The deformation of spacetime means that time as well as space is altered.) Seen from inside the black hole, everything falls dramatically and nemesis awaits only moment away; seen from outside, nothing changes.

We may think of a black hole as a region into which space flows inward at the speed of light from the outside world. Most black holes probably rotate, and we should therefore imagine inflowing space as swirling around, like water draining away in the sink. To us, accustomed to thinking of space as little more than a vacuity, this seems an incredible way of visualizing a black hole. Remember, however, that space is no longer an empty nothing.

Light outside the black hole travels through infalling space and can escape. At the surface, however, space falls at the speed of light, and rays of light seeking to escape through infalling space remain stationary. This is the event horizon. It is the country of the Red Queen of *Alice in Wonderland.* "Now, here, you see," said the Red Queen to Alice, "it takes all the running you can do, to keep in the same place." Space inside the black hole falls even faster than the speed of light, dragging everything with it, including light, and nothing can possibly escape.

Far from the black hole space is almost flat and practically the same as in special relativity. Near the black hole space is greatly curved and very much deformed. As we approach the black hole the curvature of space increases and we see less and less of what lies ahead. At the last moment before being engulfed we see only what lies behind in the outside world and nothing of what lies ahead. Nature takes pity on us and veils from view our doom.

If by mischance we stumble into a massive black hole of low density we might quite easily not realize that anything untoward has happened; only slowly would it dawn on us that we are caught in the grip of a black hole and that an awful unseen fate lies ahead.

*   *   *

Stars in their death throes are thought to be the birthplaces of black holes. When a star has consumed its central supply of hydrogen, it swells up and becomes a red giant. The Sun in about five billion years will also become a red giant. Its core or central region will contract and at the same time its mantle (outer region) will expand and fill the sky, engulfing Mercury and Venus, and perhaps even the Earth. After some tens of millions of years as a red giant, the Sun will puff away its inflated mantle and reveal a condensed central core. It will then be a white dwarf having a size about that of the Earth, and will bathe the dead terrestrial surface with a pale white light scarcely brighter than present moonlight.

Stars more massive than the Sun do not give up the game so easily. After becoming red giants they convert their helium into heavier elements, carbon, oxygen, and so on, all the way to iron, unlocking more and more nuclear energy. Some of this energy spills over and is used to manufacture elements heavier than iron, such as gold and uranium. The star has now become fiercely bright, demanding more and more nuclear energy, and its diminishing reserves are soon exhausted. Only gravitational energy finally remains, and to draw on this supply of energy the core must continue to shrink, progressively getting smaller, denser, and hotter. In the meantime, the core generates copious neutrinos that stream out of the star adding to the loss of energy.

The central density and temperature continue to rise and eventually reach a point where the heavy elements in the core are crushed and broken down into helium. Finally, the helium dissolves into its constituent particles: protons, neutrons, and electrons. The electrons get squeezed into the protons, leaving neutrons as the dominant survivors in the collapsed core.

For hundreds of millions of years the bright star has radiated into space an immense amount of energy obtained by burning hydrogen into helium. Now, confronted with the dissolution of helium, the star must repay all this energy almost immediately. At death's door, faced by ruinous debt, the star does the only thing possible. It draws on its reserves of gravitational energy by collapsing. It collapses catastrophically. The inrushing core terminates as a neutron star and the outrushing mantle signals the birth of a new star. The mantle of unburned hydrogen and helium explodes and a brilliant supernova briefly outshines all the stars of the Galaxy.

It is believed that supernova are also the mysterious "gamma ray bursters." (Gamma rays are high energy photons.) These are objects that emit a brief burst of gamma rays that generally last a few seconds. The energy released is immense and can be detected at high redshifts at cosmic distances. Many theories have been proposed to explain gamma ray bursters. One possibility is that a collapsing star in its last moments emits the burst of gamma rays. The core as it collapses rotates faster and becomes flattened. Neutrinos generated by high temperature and trapped by high density in the infalling core escape by forcing their way along the rotation axis and emerge as oppositely directed high-energy jets.

A fraction of the heavy elements escapes into space and intermingles with the interstellar gas from which new stars and their planets form. Look at any metal coin and ask, "Where was this metal made?" The answer: It was made in the Promethean fires of a stellar nuclear reactor that died more than five billion years ago before the birth of the Sun.

From the death of the old star a neutron star is born. One thimbleful of neutron matter would weigh a billion tons on Earth. Quite likely, the newborn neutron star is a rapidly rotating pulsar sending out a pulselike message of matter stressed to its limits. Its searchlight beam sweeps across the sky, and we observe the beam repeatedly as it periodically passes the Earth. At first the pulsar gyrates hundreds of times a second. Slowly it loses rotational energy and turns less rapidly and after millions of years it lapses into silence.

Larger stars, having more massive imploding cores, cannot terminate as neutron stars. The neutron matter in those collapsing cores more massive than three times that of the Sun is unable to withstand the intense pull of gravity. These cores continue to collapse and become black holes. The laws of nature as we understand them lead inevitably to this conclusion and there is now little doubt among astronomers that massive stars at the end of their evolution give birth to black holes.

Massive stars are often members of binary systems. These stars orbit each other, sometimes rapidly in close embrace, exchanging matter and evolving in spectacular ways. When a massive member of a binary system collapses it becomes either a neutron star or a black hole. Gas flows from the companion, spiraling in to the surface of the collapsed star, and some of the gravitational energy released is radiated away in a the form of X-rays.

The study of these powerful X-ray sources indicates in many cases that the collapsed companion is sufficiently massive to be nothing less a than a black hole. We are unable to see these black holes, but radiation emitted by infalling gas betrays their presence; they also affect strongly the motions of companion bodies, as the Reverend Mitchell foresaw.

*   *   *

A black hole voraciously consumes all that it encounters and is aptly described by the words of Jonathan Swift:

> All-devouring, all-destroying,
> Never finding full repast,
> Till I eat the world at last.

Once born, it grows and puts on weight. The surrounding gas spirals in and is sucked up. Incautious stars straying too close are torn to shreds by tidal forces and their wreckage adds to the headlong rush of "atoms and systems into ruin hurled." The inwardly spiraling gas, squeezed to high temperature, violently radiates and an appreciable fraction of the accreted mass is transformed into escaping radiant energy. A black hole on the prowl is an efficient engine that converts directly into radiation a significant fraction of the mass of whatever it devours.

The central regions of giant galaxies swarm with closely packed stars. Consider what happens when black holes are unleashed among these rich star systems: they become star destroyers, devouring everything, even one another, never finding full repast until they have eaten away the center of the galaxy. After hundreds of millions of years a black hole attains a mass possibly a billion times that a of the Sun. During its growth it pours forth a torrent of radiation perhaps more intense than all the stars of the galaxy.

Astronomers believe that the distant and brilliant quasars are massive black holes accreting matter in the nuclei of giant galaxies. According to this theory a quasar becomes quiescent when a black hole has swallowed the surrounding gas and stars. Also, when a black hole has grown extremely large, stars fall straight in without first being disrupted by tidal forces, and their stellar wreckage ceases to contribute to the inward spiraling of luminous gas.

Possibly, quasars have a lifetime of a billion or so years, and the majority of them existed shortly after the birth of galaxies. Their light has taken billions of years to reach the Earth and most quasars seen by us lie at vast distances.

*   *   *

Einstein showed how gravity can be interpreted as curved spacetime. In the dynamic picture of general relativity, however, electromagnetic forces remain much the same as before and lack a similar lucid geometric interpretation. For years Einstein endeavored to unify the electromagnetic and gravitational forces within a fully geometrized picture of the universe.

With Nathan Rosen, Einstein wrote in 1935, "In spite of its great success in various fields, the present theoretical physics is still far from being able to provide a unified foundation on which the theoretical treatment of all phenomena could be based." The basic idea of general relativity, applicable to matter on the large scale, fails to account for the atomic structure of matter and for the various quantum effects. Einstein never succeeded in his quest for a more unified picture. What he had achieved, though nearer to the heart's desire, fell a long way short of what he himself desired.

We no longer hope to geometrize in the manner of general relativity all the various field interactions of the natural world. New conceptual schemes are in the making, of grand unified theories that combine the electromagnetic, weak, and strong fields into a single protean field, and of exotic supersymmetry theories that seek to combine grand unification and gravity. Geometry is taking a new turn. One idea is that the world is spanned by ten dimensions, six of which are compacted into strings and other minute manifolds distributed in the other four dimensions of spacetime. We are at the stage of wondering to what extent these representations whose function mimics that of the observed world are actually the real world. Does the expression "real world" any longer have meaning?

In the future much will be understood that to us is now perplexing. By then almost certainly we shall have uncovered new riddles. The unknown will loom as large as before, possible more so, and the heart will yearn for revelation that when found will lead inevitably to the discovery of fresh mystery. The more we know, the more aware we become of what we do not know.

12　The Cosmic Tide

We live in the Solar System on the planet Earth that revolves with other planets around a star called the Sun. Light from the Sun hurrying at great speed takes 500 seconds to reach the Earth and five hours to reach the far-flung planet Pluto. The Earth that to us seems large is dwarfed by the Solar System with its whirling planets.

Starlight from the nearest stars travels for years before reaching the Earth. If we imagine the Sun having the size of a grain of sand, the nearby stars on the same scale would be at a distance of one hour's drive on an interstate highway. Scattered out to enormous distances in all directions are a hundred billion stars that constitute the whirlpool system called our Galaxy. The Galaxy – a glittering carousel of stars across which light takes 100,000 years to travel and around which the Sun journeys once every 200 million years – seems incomprehensibly large compared with the solar system.

Much has been discovered about the Galaxy: its many kinds of stars, sunlike stars, blue, yellow, and red giants, binary stars, white dwarfs, and dense neutron stars; its great spiral disk seen by us as the Milky Way where clouds of glowing gas and obscuring dust give birth to new stars; its even greater halo of very old stars and globular clusters; and still much that remains to be discovered.

Newton's universe of uniformly distributed stars has become Wright's universe. Beyond the Galaxy out in the depths of space lie hosts of galaxies of all kinds. The galaxies are the atoms of the universe.

Light from the nearest galaxies takes about one million years to reach the Earth and we see them as they were when early human beings gazed at the sky with wondering eyes. Most galaxies are midget systems of only tens of millions of stars. But not all. The neighboring giant galaxy of Andromeda at a distance of two million light-years, festooned with outlying midget systems, is much like our Galaxy. Scattered here and there we see giant ellipticals, also giant spirals similar to our Galaxy, and sometimes supergiant galaxies a hundred times more massive still. Among the giant galaxies can be found the powerful and still enigmatic radio sources.

Through telescopes we see the majestic galaxies stretching away like celestial cities to unlimited distances. We look out into the depths of space and see galaxies as they were billions of years ago at a time before life arose on Earth. In their midst gleam the intensely bright quasars.

The galaxies, as Kant foresaw, cluster together to form even larger systems. Clusters of galaxies come in all sizes. Our Galaxy and its great companion galaxy of Andromeda are the dominant members of a swarm called the Local Group. Most clusters are comparatively small like our Local Group and contain tens of galaxies. But richly populated systems, such as the Coma and Perseus clusters, contain galaxies by the thousand, and often in their central regions blaze the supergiant galaxies.

Galaxies are the cradles of life. Who doubts the existence of life out there in the galaxies? In many cases that life is perhaps more intelligent than that on Earth. Even if we were so begrudging as to concede that life exists on only one planet to each galaxy, then in the colossal observable universe there would still be trillions of planets inhabited by living creatures.

The last shadows of mythical anthropocentrism melt away before the astronomical grandeur of the physical universe.

* * *

The history of cosmology unfolds a growing conviction that human beings do not occupy a position of central importance in the cosmic scheme.

The assault on the mythic universe by Hellenic science, followed by the Copernican and Darwinian Revolutions, dethroned the human species. The cosmic center – first the nation, then the Earth, the Sun, and finally the Galaxy – has vanished from the physical universe. From science, with help from theology and philosophy, emerges an outlook expressed by the location principle.

The location principle states: *it is improbable that human beings have a central location in the physical universe.* Other planets encircle other stars in other galaxies in other clusters and may have life that may in many instances be more advanced and perhaps more precious than on Earth. Why then should human beings be singled out for special location? The location principle states that of all the planets, stars, galaxies, and clusters in the universe it is improbable that the Earth, Sun, Galaxy, and Local Group are in any way uniquely privileged. We can be kings of the cosmic castle in the mythic universe but not in the modern physical universe.

*　　*　　*

The sea, seen from a ship, stretches away the same in all directions. The waves disturbing the surface are no more than incidental irregularities. Lucretius echoed the Atomists when he said, "the universe stretches away just the same in all directions without limit." Yet this ideal in the minds of the Atomists is not in the least obvious to an observer looking out from Earth.

We are afloat, it seems, in a cosmic ocean surrounded by great waves that at first glance do not appear in the least like incidental irregularities. Only when we look out far beyond the Galaxy do we find that on average things in one direction look much the same as those in other directions. All directions look alike, and from our particular viewpoint the universe is isotropic.

Discussions in modern cosmology place considerable weight on what is now called the *cosmological principle.* The cosmological principle, so-named by the astrophysicist Edward Milne in 1933, was expressed by him in the words: "Not only the laws of nature, but also the events occurring in nature, the world itself, must appear the same to all observers, wherever they may be." The principle states that when local irregularities are ignored, or averaged out, the universe at any instant in time is the same everywhere in space. As Einstein said in 1931, "all places in the universe are alike." In other words, the cosmological principle asserts that the universe is fundamentally homogeneous in space.

The concept of cosmological homogeneity originated with Anaxagoras and the Atomists, and later reemerged in the late Middle Ages. Cardinal Nicholas of Cusa in the fifteenth century, using the analogy that God is ubiquitous and uncircumscribed, declared "the fabric of the world has its center everywhere and its circumference nowhere." What is potential in God is made actual in the created universe.

Nowadays, astronomers observe isotropy (all directions are alike) and theoretical cosmologists postulate homogeneity (all places are alike). The probability argument of the location principle links together observed isotropy and postulated homogeneity.

Let us imagine that we stand on the summit of a hill from which the surrounding landscape looks much the same in all directions. The scenery from our vantage point appears isotropic. But we are not at liberty to declare that all places are alike and the landscape is homogeneous. When seen from any other point of view, the scenery is not the same in all directions. Our isotropic view from the summit is the consequence of a central location.

Similarly, if we occupy the center of the universe, as in the Aristotelian or medieval universes, our central location explains why the fixed stars in all directions appear much the same.

Unfortunately, we are confined to a small region of the universe and cannot travel elsewhere to another vantage point – say a few thousand million light years away – to take a fresh look at the cosmic scenery. Instead, we must use the location principle that assures us that central location is improbable. From this principle we draw the conclusion that all directions are alike, not only from our place in the universe, but from all other places. Isotropy is not unique to our place but is the same everywhere.

It comes to this: when irregularities are ignored, the observation that the universe is isotropic from our place in space, coupled with the location principle that says central location in the universe is improbable, leads to the conclusion that the universe probably is isotropic from all other places and consequently is homogeneous.

We must imagine that we stand not at the summit of a hill but on a flat or spherical surface, and all directions are alike at every place simply because the surface is everywhere the same. Such a surface is homogeneous. Analogously, the universe is homogeneous, as asserted by the cosmological principle.

Apart from astronomical irregularities (planets, stars, galaxies, clusters), cosmic space is either flat or uniformly curved. With cosmic space comes *cosmic time*, ticking away everywhere at the same rate. If we could rush around the universe at infinite speed, we would find at any instant that everywhere seems much the same, the laws of nature the same, clocks running in synchronism, and things evolving in similar ways.

The cosmological principle – founded on astronomical observations and a probability argument – unites the universe into a homogeneous whole.

*   *   *

The expanding universe ranks among the most startling discoveries made in the twentieth century. The galaxies are drifting apart and the yawning spaces between are widening.

From the speed at which the galaxies move apart we can estimate that long ago, somewhere between 10 and 20 billion years, everything existed in a state of extreme congestion commonly referred to as the big bang.

In recent years a second startling discovery has been made. The afterglow of the big bang suffuses the whole of space. This ubiquitous glow, invisible to the unaided eye, is the three-degree cosmic radiation discovered by Arno Penzias and Robert Wilson in 1965. Its rays travel freely in space, coming to us in all directions, and the extraordinary isotropy of this radiation reinforces our belief in the basic homogeneity of the universe. Although of very low temperature (three degrees above absolute zero, which is 273 degrees below the freezing point of water), this radiation ranks as a highly important constituent of the universe. Hold up the palm of your hand to the sky, day or night, and a thousand trillion photons – particles of light – of the cosmic radiation will strike it in one second. This radiation, now cooled and enfeebled by expansion, long ago was the incandescent light of the early universe.

The Stoics believed in a universe of periodic fiery explosions and implosions. Edgar Allan Poe in his imaginative essay *Eureka* of 1848 vividly portrayed the possibility of a pulsating universe, expanding and collapsing: "Are we not, indeed, more than justified in entertaining the belief – let us say, rather, in indulging a hope – that the processes we have ventured to contemplate will be renewed forever, and forever, and forever; a novel universe swelling into existence, and then subsiding into nothingness, at every throb of the Heart Divine." The idea of a cyclic universe expanding and collapsing, bouncing from big bang to big bang, each a throb of the heart divine, each a day in the life of Brahma, still persists to this day.

Edwin Hubble's observations in the late 1920s and early 1930s, with contributions by other astronomers, have made secure the idea of an expanding universe. Often it is too troublesome to trace an idea or a discovery back to its origin and it is more convenient to make attribution to the one who convinced the world. As the New England poet James Lowell said,

Though old the thought and oft expresst,
'Tis his at last who says it best.

Correctly or incorrectly we attribute to Hubble the discovery of the expansion of the universe.

*   *   *

The story begins in 1912 with Vesto Slipher of the Lowell Observatory measuring the shift in the spectral lines of the light emitted by other galaxies. From his measurements of spectral shifts he calculated the velocities at which these galaxies are approaching and receding. By 1923, as a result of Slipher's painstaking work, it was known that of the forty-one galaxies studied, five are approaching and thirty-six receding. Clearly, if galaxies moved randomly in all directions, about half should be approaching and half receding. Slipher's observations showed that the galaxies had a mysterious tendency to recede.

On the front page of *The New York Times* in 1921, under the heading 'Celestial Speed Champion', Slipher reported his latest discovery of a receding galaxy:

> The lines in its spectrum are greatly shifted, showing that the nebula is flying away from our region of space with a marvelous velocity of 1100 miles per second... If the above swiftly moving nebula be assumed to have left the region of the sun at the beginning of the earth, it is easily computed, assuming the geologist's recent estimate of the earth's age, that the nebula now must be many millions of light years distant. The velocity of this nebula... further swells the dimensions of the known universe.

The distant galaxies (or nebulae) are running away and in the past must therefore have been much closer together.

Albert Einstein and the Dutch astronomer Willem de Sitter proposed in 1917 quite different models of the physical universe based on the theory of general relativity. Einstein's version had spherical geometry and contained matter; it was closed, and a person traveling in a straight line for a long period of time would eventually return to the starting point from the opposite direction; moreover, it was static, neither expanding nor collapsing. The de Sitter version contained no matter; it was open, and space extended to infinite distance in all directions and could not be circumnavigated.

The de Sitter universe with its pathological absence of matter might have been ignored and forgotten but for one particularly interesting feature. Particles of matter when sprinkled in it shared a tendency to move apart. Cosmologists conjectured that this *de Sitter effect* might have some bearing on the results obtained by Slipher. A few years later cosmologists realized that the de Sitter universe was expanding. An apt distinction was then drawn: the Einstein universe consisted of "matter without motion," and the de Sitter universe consisted of "motion without matter."

The German astronomer Carl Wurtz, prompted by Slipher's velocity measurements, proposed in 1922 a sort of velocity–distance law, according to which the farther away a galaxy, the faster it recedes. He estimated the distances of galaxies by their apparent sizes and made the not very reliable assumption that the smaller the apparent size of a galaxy the greater its distance.

Also in 1922 the Russian physicist Alexander Friedmann published a paper 'On the curvature of space' in a distinguished German scientific journal. In this pioneer work he investigated "non stationary worlds" that either continually expand or first expand and later collapse. A second paper followed in 1924 in which Friedmann looked at other nonstatic worlds. Both papers drew little attention. Georges Lemaître, a Belgian mathematician and an ordained priest, published in 1927 a paper entitled 'A homogeneous universe . . . accounting for the radial velocity of extra-galactic nebulae' in which he explored many of the characteristics of expanding universes. This work also drew little attention until translated into English four years later in the *Monthly Notices* of the Royal Astronomical Society of Great Britain.

Hubble meanwhile had undertaken the task of measuring the distances of nearby galaxies using special pulsating stars of known intrinsic brightness called cepheids. In 1928, Howard Robertson used Slipher's velocity measurements and Hubble's distance estimates to derive a linear velocity–distance law. This law was firmly established in 1929 by Hubble, and the growing volume of information showed clearly that the galaxies are receding and the universe is expanding.

In his book *The Expanding Universe* Arthur Eddington wrote in 1933: "The unanimity with which the galaxies are running away looks as though they had a pointed aversion to us. We wonder why we should be shunned as though our system were a plague spot in the universe. But this is too hasty an inference and there is really no reason to think that the animus is especially directed against our galaxy." To youths such as myself Eddington explained that the galaxies are not running away from us, but from one another. Astronomers in all galaxies have at first the impression that their particular galaxy is the plague spot of the universe.

*   *   *

We come now to an important point. The universe does not expand in space but consists of expanding space. The galaxies do not hurtle away through space. Such a view belongs to the archives of the Newtonian universe. Space has now become an active participant on the cosmic stage. The galaxies float at rest and are carried apart by the expansion of space.

Of course, the galaxies are never exactly at rest in expanding space. They have their local and random motions, usually within clusters, and this explains why some of the nearest galaxies are approaching and not receding. Also, most clusters do not expand, and only the space between them expands. We are only interested in the main outline, however, and for simplicity we shall think of the galaxies as unclustered and at rest in expanding space.

To make the picture clearer, let us consider an expanding sheet of rubber. We imagine that this flat surface represents the space of our physical universe. The sheet expands uniformly. By this I mean it expands isotropically (the same in all directions) and homogeneously (the same at all places). A triangle drawn anywhere on the expanding surface remains a similar triangle while its size steadily increases.

We draw a circle and declare that it represents a galaxy. But as the surface expands, this "galaxy" gets bigger. A real galaxy, held together by its own gravity, is not free to expand with the universe. If the circle is labeled star, or planet, or atom, this also would be wrong, because all such objects are held together tightly and are not free to partake in the cosmic dilation. We observe the expansion of the universe because our observatories and measuring instruments have fixed sizes.

We erase the circle and replace it with a small paper disk. As the surface expands, the disk remains constant in size, and hence we have found a way of representing a galaxy in an expanding universe.

Over the surface, more or less uniformly, we scatter a large number of paper disks of various sizes. They remain at rest on the expanding surface and retain their fixed sizes. From each disk the surrounding disks recede. Imaginary inhabitants on any one disk have the impression that they occupy the cosmic center from which everything is running away. But the inhabitants on all disks share this impression and there is no actual center.

It is easy to demonstrate the velocity–distance law. We choose any disk and label it A. A second disk, labeled B, at a certain distance from A moves away at a certain velocity. A third disk, labeled C, in the same direction as B and at twice the distance moves away from A at twice the velocity of B. Obviously, by virtue of homogeneity, C must move away from B at the velocity that B moves away from A. The easiest way to demonstrate this is with a length of elastic that has attached paper clips spaced at equal intervals. As the elastic is stretched we see the paper clips move away from one another at relative velocities proportional to their spacings. The physical universe behaves in much the same way, and the farther apart the galaxies, the faster they recede from one another.

*   *   *

The velocity–distance law seems simple enough to us looking down on the expanding sheet, like gods surveying the Trojan Plain. We see what happens everywhere in cosmic space at a fixed instant of cosmic time. But the inhabitants on the disks find the situation far from simple, and the terms velocity and distance used in the velocity–distance law need careful interpretation.

The disks rest on the surface and move apart because the surface expands. Relative to one another they have recession velocities. The galaxies at rest in expanding space also have recession velocities relative to one another. But in addition to their recession motion galaxies move to and fro inside their clusters and have peculiar motion. They have a peculiar velocity as well as recession. Peculiar motion – of local importance but not of great cosmological significance – applies to bodies that move through space. This ordinary motion, familiar to us on Earth, in the Solar System, and in the Galaxy, is subject to the rules of special relativity; it never exceeds the speed of light.

Recession velocity applies to motion produced by the expansion of space and is exempt from the rules of special relativity but not general relativity. This is the expanding space paradigm that cosmologists grappled with in the 1930s. Failure to distinguish between peculiar velocity and recession velocity leads to confusion for the beginner in modern cosmology. It is not the peculiar velocity of bodies moving through space (which is familiar), but about the recession velocity of bodies comoving with space (which is unfamiliar) that must be used in the velocity–distance law.

The measurement of distance with a tape measure is no great problem for us who gaze down on the expanding surface. But from our godlike eminence, even we must exercise a little care and ensure that all distances in the velocity–distance law are measured at the same instant in time. When a disk at a distance of 1 meter from a chosen point recedes at 1 centimeter a second, we note that another disk at a distance of 10 meters recedes at 10 centimeters a second. Both distances are of the tape-measure kind and are measured simultaneously. All distances must be determined at the same instant, because the surface may not be expanding at constant rate. The expansion of the rubber sheet may be speeding up or slowing down. Similarly, the universe may not be expanding at constant rate.

Measuring the distances of remote galaxies is an arduous undertaking. First we must know the distance to the Sun; then by means of parallax measurements we find the distances of nearby stars. By comparing stars of known brightness we find the distance of star clusters farther away, and by comparing star clusters we reach out to greater distances. The bright cepheid variable stars acquire in this way known intrinsic brightnesses and play an important role as indicators of distance. Patient work with careful interpretation determines the distances of star clusters and cepheid variables in the nearest galaxies. For galaxies farther away we use whatever suitable distance indicators are available, such as the brightest stars, luminous clouds of gas, and certain kinds of supernovas. Very luminous galaxies of estimated intrinsic brightness then become bright beacons that help to determine the distances of rich clusters, some so distant that small groups like the Local Group are undetectable.

We see the galaxies as they were in the past, long ago, and allowance must be made for their evolution and change in brightness before comparing them with nearby galaxies. The whole subject of distance measurements in astronomy is an intricate art in which uncertainties unavoidably increase with distance.

We see the galaxies not where they are now but where they were when the light we see was emitted. The estimated distances must be adjusted to a common instant of cosmic time, say the present epoch, before using them in the velocity–distance law. Estimating the distances of galaxies is difficult enough, adjusting these distances to a common epoch is even more difficult. Adjustment requires that we know not only how the galaxies evolve, but also how the expansion of the universe changes with time.

For every million light-years of distance the recession velocity increases by about 20 kilometers a second. This expansion rate is ten times less than that first estimated by Hubble with his limited information and rough-hewn estimates of distance, and even now it is still uncertain by perhaps as much as a factor of two.

Before leaving the model we perform one more experiment. We arrange first that the expansion occurs at a constant rate. We then sprinkle fresh disks on the expanding surface at a constant rate in such a way that nothing ever seems to change. New disks occupy the widening gaps between old disks and the average separating distance between the disks remains unchanged. This illustrates what happens in the steady-state universe that was proposed in 1948 by Hermann Bondi, Thomas Gold, and Fred Hoyle: new galaxies form continually from freshly created matter and the cosmic scenery remains permanently unchanged. All places on average are alike in time as well as in space.

We know that for every million light-years of distance the recession velocity increases by about 20 kilometers a second. If we divide 1 million light years by 20 kilometers a second, we get 15 billion years, which is a crude estimate of the age of the universe. In the 1920s and 1930s the recession velocity was thought to be ten times larger, thus giving an estimated age of from 1 to 2 billion years. This result, implying a universe that was younger than the Earth, created a cosmological paradox. The steady-state universe proposed in the late 1940s was an ingenious way of circumventing the paradox. This bold theory provoked considerable controversy, echoing the heated debates between the catastrophists (now the big-bangers) and the uniformitarians (now the steady-staters) in the early nineteenth century. A self-replicating universe of continuous creation, eternally unchanging in appearance, though fascinating to some was repugnant to others.

More precise measurements have since increased the estimated age of the universe and it is now reckoned to be sufficient to accommodate the oldest stars. The original main purpose of the steady-state universe no longer exists.

The three-degree cosmic radiation discovered in 1965 indicates that the universe was once exceedingly dense and hot, and it is safe to say the physical universe cannot be eternally unchanging in appearance, as supposed by the advocates of the steady state theory.

*   *   *

The recession velocity, according to the velocity–distance law, increases steadily with increasing extragalactic distance. Eventually we reach the edge of the *Hubble sphere*. At a distance of roughly 15 billion light-years – the radius of the Hubble sphere – the recession velocity matches the velocity of light. Inside the Hubble sphere things recede slower than the velocity of light; outside the Hubble sphere things recede faster than the velocity of light.

How is it possible for anything outside the Hubble sphere to move away faster than light? The answer is the expanding space paradigm. Nothing moves through space faster than light, a feature of special relativity; but space itself, however, has dynamic properties governed by general relativity, and can expand faster than light.

We know that the universe has no edge and cannot terminate abruptly. Either space extends to infinity or curves back on itself like the surface of a globe. Homogeneity in both cases requires that the recession velocity progressively increases and eventually exceeds the speed of light outside the Hubble sphere.

There are as many Hubble spheres as galaxies. Each galaxy has its own Hubble sphere at the surface of which the recession velocity from the galaxy equals the velocity of light. The cosmic edge cannot exist at the surface of our Hubble sphere because all galaxies have their own Hubble spheres and the universe would have as many cosmic edges as galaxies.

Suppose for a moment that the surface of our Hubble sphere were indeed the cosmic edge. Nothing now exists outside and hence nothing recedes faster than the velocity of light. Those unable to accept the expanding space paradigm can breathe a sigh of relief. But at what a cost! Other galaxies are denied similar Hubble spheres and condemned to a lopsided view of the universe. All places are no longer alike. We have restored our privileged position at the center of the universe, and by throwing away the location principle have lost assurance that things out there obey laws similar to those here. We might as well give up cosmology as a subject of scientific inquiry.

All galaxies stand on equal footing and have their own Hubble spheres. This means that things exist outside our Hubble sphere and recede from us faster than the velocity of light. The expanding space paradigm lies at the heart of modern cosmology.

*   *   *

Light travels at constant speed measured locally in the space through which it travels. Nothing in nature has an ordinary speed exceeding the speed limit of light. All ordinary or peculiar velocities are subject to this limit, but recession velocities are without limit.

The velocity–distance law tells us that the greater the distance the greater the recession velocity. At infinite distance in an open universe the recession velocity is infinitely great. Recession occurs because of the expansion of space and does not consist of ordinary motion through space. Those persons who find it difficult to understand how recession can be without limit generally make the mistake of supposing that the galaxies are shooting away through space like projectiles. They have failed to realize that the galaxies are at rest in expanding space.

Consider a galaxy outside our Hubble sphere. Light rays from the galaxy, emitted in our direction, hurry toward us and travel through space that recedes faster than the speed of light. Thus even the light emitted by the galaxy recedes from us. As Arthur Eddington said in *The Expanding Universe,* "light is like a runner on an expanding track with the winning-post receding faster than he can run."

A galaxy at the edge of the Hubble sphere recedes at the speed of light. Its rays emitted in our direction stand still relative to us. Here again is the country of the Red Queen where however fast Alice runs she remains stationary and goes nowhere.

Outside the Hubble sphere even light is receding. It might be a mistake, however, to suppose that our Galaxy will never in the future receive this light. In a decelerating universe (the rate of expansion is slowing) the Hubble sphere itself expands generally faster than the universe and its surface sweeps out and overtakes the distant galaxies. In this sort of universe the inflating Hubble sphere progressively contains more of the universe, and the number of galaxies inside increases by approximately ten each year. A galaxy outside the Hubble sphere in a decelerating universe may one day be overtaken; it will then lie inside, and its emitted light rays at last will be able to approach our Galaxy and be received. Eddington's runner must not give up the race but keep on running, because the expanding track may be slowing down, and the winning post will eventually be reached. Recent observations indicate, however, that the universe may in fact have entered a stage of acceleration, and in that case the Hubble sphere may not be expanding but is more or less fixed in size. A galaxy outside the Hubble sphere will therefore never be seen; the winning post will always recede faster than the runner can run.

We cannot see the whole universe, only that part around us referred to as the observable universe. There are a few technical complications concerning the horizon of the observable universe that need not bother us in this discussion. In a decelerating universe the observable domain increases in size in much the same way as the Hubble sphere, and in the course of time we see more and more of the distant universe. In a universe that first decelerates and then begins to accelerate, as is now currently thought, the observable domain stays roughly constantly in size and in the course of time the galaxies recede out of sight and we see less and less of the contents of the universe.

*   *   *

We know that the universe expands because the light received from distant galaxies is redshifted. Red light has longer wavelengths than blue light. The light emitted long ago by a distant galaxy is received by us as red light. The light is not reddened by the removal of blue light, as in a fog, but all wavelengths are stretched or redshifted. Light rays journey for long periods of time over vast regions of expanding space

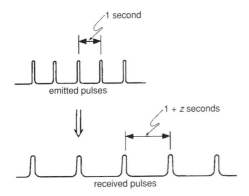

A distant galaxy emits a pulse of radiation once every second. The space between the pulses is stretched while the pulses travel through expanding space and they are received in our galaxy every $1 + z$ seconds, where z is the redshift.

and what happens is easy to understand. The rays are stretched by the expanding space through which they travel. All wavelengths are stretched and blue light slowly changes into red light. Take a length of stiff wire and bend it into a wavy or snakelike shape; now slowly pull on both ends and notice how the waves get longer. This is analogous to what happens to waves of light traveling through expanding space. The light received is redder than the light emitted.

A galaxy emits rays of light that eventually are observed in another galaxy far away. While the rays travel through expanding space between the two galaxies, their wavelengths steadily increase. Finally, the rays enter a telescope in the receiving galaxy and the astronomers notice that all wavelengths have increased by a certain amount. They study its spectrum, comparing it with the spectra of luminous sources in their own galaxy, and in this way determine the amount of the redshift. The astronomers assume that the light-emitting atoms in their own galaxy are similar to the light-emitting atoms in the distant galaxy. They, in fact, assume that the universe is homogeneous in the sense that atoms and the laws of nature are everywhere the same. To justify this far-reaching assumption they observe that the galaxies in all directions appear much alike, and by invoking the location principle they deduce homogeneity.

Let us suppose the astronomers discover that the received rays have their wavelengths increased twofold. From this amount of redshift they know that the universe has expanded twofold since the rays were emitted. During that time the average density of matter in the universe has decreased eightfold.

Expansion redshifts are extremely useful. They tell us directly how much the universe has expanded during the time between emission and reception of light. They do not tell astronomers how fast the galaxies recede, or how far away they are, or how long ago they emitted the light now received; this information must be deduced within the framework of a theoretical model. Instead, they tell astronomers precisely how much the universe has expanded between emission and reception of extragalactic light.

Astronomers in the 1920s observed that extragalactic redshifts increased with distance; the farther away a galaxy, the greater its redshift. A linear redshift–distance relationship, in which redshift increases with distance, is known as Hubble's law. To a first approximation Hubble's redshift–distance law is the same as the velocity–distance law. But only for small redshifts. The velocity–distance law holds for all distances, but the redshift–distance law holds only for distances much less than about one billion light-years.

Let us suppose that a distant galaxy emits a pulse of light once every second, and we in our Galaxy detect these pulses of light. The pulses on leaving the emitting galaxy have initially a separation in space of one light-second. While they travel through the intervening expanding space their separation increases. The pulses are not received by us at a rate of one every second but at a slower rate, because their separation in space is now increased. When the redshift tells us that wavelengths have been stretched twofold, the pulses arrive in our Galaxy with a separation of two light-seconds, and are received at two-second intervals. If the pulses are emitted once every year as measured by a clock in the emitting galaxy, they are received once every two years as measured by an identical clock in our Galaxy. A person in the distant galaxy living for three score and ten years appears to us to live for seven score years.

At large redshifts things seem to change more slowly than locally. Out near the horizon of the observable universe, where all

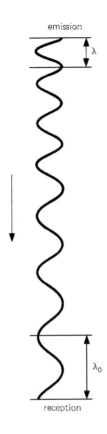

emission

A wave of radiation is stretched and its wavelengths are increased as it travels in expanding space. If λ is the emitted wavelength, the received wavelength is $\lambda_0 = \lambda(1 + z)$, where z is the redshift.

reception

appears extremely redshifted, time has slowed down to a snail's pace, or so it seems to us here.

*   *   *

In some popular treatments of modern astronomy and cosmology the reader gains the impression that the physical universe is a world of extreme violence where the galaxies shoot away through space like the ejecta of an immense explosion and are themselves the scenes of violent cataclysms.

If we must be anthropocentric, then the people of the high Middle Ages, with their universe of harmonious spheres, were perhaps closer to the truth. The stately orbs and their measured tread in unison to the music of the spheres have in the modern universe become the galactic cities, alit with starlight, drifting serenely in the cosmic tide. The music of the spheres has become the music of space and time and the symphony of harmonious forces. Once, long ago, it all began in an age of brilliant light. The closer we examine the design of the physical universe the more we marvel at its harmony and its fitness for habitation by life.

13 Do Dreams Come True?

Historians would love to search the past in a Wellsian time machine and return to tell the "tales of long, long ago, long, long ago" that in the words of Thomas Bayly, a nineteenth century ballad writer, "to us are so dear." Historians little know that a timeship has been invented by a professor in the Department of Fantasy and Virtual Reality at the University of Massachusetts. In this secret diachronic conveyance we shall take a journey – a safari in time – back to earlier periods of cosmic history.

Let me welcome you aboard with these comments. Moving backward in time is an uncommon way of presenting history, and to avoid the incongruity of a movie show in reverse, I shall occasionally stop the machine and allow time to resume its normal Newtonian flow while gazing at the scenery. I must warn you that our timeship is still in an experimental stage and will not always do exactly what we want. Please fasten your seat belts.

Tentatively I start the timeship in reverse gear and it lurches into motion. Its dials spin alarmingly, and although I slam on the brakes almost immediately, we have already traveled two million years. Through the windows we see hominids striding around in the early Pleistocene. It would be very interesting to stay and see their progress. But we have other more urgent business.

We move on and at ten million years we again stop briefly and see the hominoids before they evolve into apes and humans. Nothing in the sky has changed – the stars and galaxies look much the same – yet we cannot help noticing how the continents are adrift on the Earth's surface. At twenty million years the primates first appear, and about this time India slams into Asia and thrusts up the Himalayan ramparts.

On we go, across the Cenozoic* era, skipping back in time some sixty-five million years. We arrive in time to witness the rise of the mammals, the emergence of grasses and flowering plants, and have unfortunately just missed the demise of the dinosaurs. We cross the Mesozoic era, pausing in the Jurassic period to photograph a few dinosaurs, and after a journey of two hundred million years we reach the Paleozoic era in time for lunch. The land masses have temporarily fused together forming the supercontinents of Laurasia to the north and Gondwanaland to the south, separated by the Tethys Sea.

After a stroll in the exotic forests and a glance around at the enormous inland seas and marshes teeming with amphibia, fish, and insects we climb aboard and recommence our journey. With its dials whirling the timeship leaps across the Paleozoic era to a time six hundred million years before the present. The great forests and all the reptiles, fish, and insects have gone, leaving an unfamiliar world to metazoans and other invertebrates.

We enter the long Proterozoic era. At one billion years before the present an unrecognizable Earth swarms with unicellular forms of life. The cell is king. After billions of year of evolution – the invention of the membrane, sex, and cell division – it stands ready with multiple specializations to form multicellular structures. Overhead the Sun pours down its unfiltered ultraviolet rays. Of course, if the news leaked out about our timeship, savants of every kind would flock to Massachusetts, queuing up to find solutions to all sorts of problems. As yet the news has not leaked out.

Our journey backward in time must now proceed at a less leisurely rate. We accelerate to greater speed. While the eons roll by we notice the distant galaxies are creeping closer. Constellations of newborn stars twinkle in the night sky and multitudes of dying stars flare up and fade into extinction. Conceivably, life originates in myriads of solar systems and perhaps in most it never evolves very far.

---

* *Ceno* means recent, *paleo* ancient, and *protero* earliest.

After a total journey of five billion years, at a time when the Galaxy is roughly half its present age, we stop for dinner and a night's rest. The evening's entertainment consists of watching the birth of the Solar System.

*   *   *

Far from the center of the Galaxy in a large interstellar cloud lurks a smaller, denser, darker, cooler region of churning gas and dust that is slowly contracting. As the dark region shrinks it swirls more rapidly. After millions of years its center grows dense and hot and develops into an embryonic Sun. Meanwhile, far from the center, colliding grains of dust coagulate to form meteoroidal rocks and ice that settle and form into an encircling disk. The rocks grow by accreting one another and form planetesimal bodies that eventually develop into the planets and their moons. The primordial Sun brightens and in a flurry of convulsive vigor thrusts back into space the remaining uncondensed gas.

The planets and their satellites then sweep up much of the flotsam and jetsam littering interplanetary space. Evidence of this period of bombardment, lasting hundreds of millions of years, is still visible on the cratered face of the Moon, and on the surfaces of planets such as Mercury and Mars.

Here is an opportunity that cannot be missed. We land on Earth and with the timeship set in forward gear we probe the future in quest of the origin of terrestrial life.

The Earth spins rapidly. Through its veils of dust we see the Sun careering across the sky with sunset following sunrise every two hours. The sky reflects the ruddy glare of lava flows, and the primitive atmosphere, fed by fiery volcanic plumes, tormented by monster storms, consists mostly of nitrogen, ammonia, methane, water vapor, and carbon monoxide, with a trace of oxygen. The smoking and steaming surface, stalked by giant tornadoes, heaves incessantly from earthquakes and the impact of huge meteorites. Piercing rays of sunlight and constant lightning conspire to establish a worldwide biochemical industry that manufactures vast quantities of an array of organic compounds, including amino acids and nucleotides.

High mountain ranges and deep oceans are landscape features of the distant future. But shallow seas abound everywhere, each a laboratory pool of primeval broth, continually boiled, shaken, and decanted. Countless myriads of biochemical experiments are performed in the seas and atmosphere every second. After hundreds of millions of years the nucleotides form into chainlike molecules of various codings, governing the assembly of amino acids into proteins of numerous kinds.

At some stage the inevitable happens: a quixotic molecule becomes self-replicating and begets a protean species whose basic design endures beyond the lifespan of its individual members. We cannot see clearly through the fog and torrential rain and we must infer that an entire sea becomes dominated by a dynasty of replicating molecules. Possibly, and here the gloom seems thicker, many seas discover their own species of replicating molecules. The living seas compete, exchanging their genetic codings by inundations, interconnecting streams, and windborne foam.

Half a billion years or so later the cells emerge in simplest form. The dilution of the seas, with water brought from the Earth's interior by volcanoes, encourages the development of replicating molecular systems that retain their own rich environment of organic compounds. The invention of membranes, enclosing tiny autonomous organic worlds, ushers in the cells.

Cells are the smallest but possibly not the first forms of life. That honor goes conceivably to the *thalassabionts* – if I may coin a word – that are the living seas of replicating molecules. Who knows whether the thalassabionts succeeded in creating a coordinated sea-wide membraneless structure? Here are the ingredients of a story to outrival all science fiction.

The cells thrive, enfolding one another and forming complex unicellular structures. Constant experimentation throughout the four billion years of the Proterozoic era is supervised by natural selection, and by trial and error the cells develop sexual and asexual division and evolve into miracles of intricacy.

*   *   *

We have detoured much too far. Reluctantly, we reverse our machine and leap back to a period before the formation of the Solar System.

After a night's rest we embark once more on our journey. While the dials of the timeship register intervals of billions of years, we observe the great clusters of galaxies slowly approaching and then merging into one another. The galaxies, released from the embrace of their dissolving clusters, drift closer and closer together, getting younger and younger.

Then, when the universe is a billion years old, the galaxies, one by one, here and there, swell up and become huge gaseous globes consisting almost entirely of hydrogen and helium. We stop the timeship and with time flowing in its normal manner we survey the scene with speculative eye.

Gradually the globes shrink and turn into bright lanterns. Their inner regions light up with swarms of first-born stars. Infalling gas descends between the stars, condensing and forming new generations of stars. In some globes the swirling gas settles and forms the rotating disks of spiral galaxies. In other globes the gas continues to fall and settles finally into the nuclei of giant elliptical galaxies. Said Hilaire Belloc in *More Beasts for Worse Children*,

> Oh! let us never, never doubt
> What nobody is sure about!

I forgot to say at the beginning that this journey occurs in the imagination, and consequently we see only what is known or thought to be known. Our timeship is really a dreamship. How galaxies form and why they exist we do not know, and whatever we say on what nobody is sure about remains purely conjectural.

Of this only we can be sure: in a heroic age the gods of morning beget the galaxies whose starlit worlds create and nurture organic life.

*   *   *

"Now entertain conjecture of a time when creeping murmur and the poring dark fills the wide vessel of the universe," said the Bard. Across a "dark backward and abysm of time" we continue our journey. Back through a turbulent darkness of the mesocosmic era to an age when the universe is tens of millions of years old. Little is known of this dark age, this strange prenatal era of the galaxies. If more were known, we might understand how galaxies form in an expanding universe.

Perhaps flickers of eerie light pierce the darkness and perhaps gas writhes in the grip of tortuous magnetic fields; perhaps strewn everywhere are black holes of assorted masses forged in the paleocosmos. Perhaps . . . Who knows?

The temperature slowly rises and the darkness melts into a faint glimmer of dull red light. At an age of one million years the universe glows red. "What dreadful hot weather we have! It keeps me in a continual state of inelegance," says Jane Austen, quoting from one of her letters. Her remark reminds me that I have forgotten to put on the air conditioning unit.

We must now proceed with circumspection and take shorter and shorter flights in time. At an age of 300,000 years the universe is filled with yellow light almost as bright as the surface of the Sun. The density of matter has risen to about a thousand atoms per thimbleful of volume, and though this may not seem much, it is more than a billion times the average density of the present universe and a thousand times more than the average density of our Galaxy.

At last we stand at the threshold of the early universe. (Or, if you prefer, the big bang.) From this epoch descends the cosmic radiation that, cooled and enfeebled by expansion, we now observe as the three-degree radiation. The study of the ripples in this radiation is now a major branch of cosmology, yielding clues on the origin of structure in the universe.

*   *   *

In trepidation we cross the threshold and enter the dreaded big bang. Much to our surprise we find ourselves in a silent and serene world of incandescent light. Like the ancient mariner we are the first that ever burst into this silent sea. We voyage across the comparatively long radiation era of the early universe, across an era that begins when the universe has an age of one second and ends when it has an age of 300,000 years. The dominant constituent in this era of cosmic youth is radiation.

We owe to George Gamow and his colleagues Ralph Alpher and Robert Herman the inspired idea of an early period dominated by pure radiation. Their theory of the radiation era, first advanced in the late 1940s, was confirmed by the discovery in 1965 of the low-temperature cosmic background radiation.

As we proceed back through the radiation era the temperature rises and the fiercely bright light soars in intensity. Glaring yellow light changes into intense white light that changes into even more intense ultraviolet light. When the temperature reaches one billion degrees, at a cosmic age of three minutes, and the light has transformed into a dense ocean of energetic X-rays, the universe becomes a nuclear reactor. A quarter of all matter in the form of protons and neutrons transforms into helium nuclei, liberating more energy than all the stars shining since the beginning. But the energy released in this thermonuclear detonation falls a long way short of that already present and the effect is only slight.

All the multitudes of stars busily converting their hydrogen into helium over ten billion years have contributed only about one-tenth of the total amount of helium in the universe. The remaining helium forms in the earliest stages of the radiation era. From the first three minutes comes also the deuterium that one day will be vitally important to us earthlings when scientists discover the trick of releasing controlled energy by means of fusion.

The radiation era commences at about the time when the universe has an age one second, a temperature of ten billion degrees, and a total density of about one million times that of water (one ton per thimbleful). We now leave the radiation era and enter the outlandish lepton era.

Lightweight particles suddenly fill the wide vessel of the universe. These newcomers generated by the intense radiation are the leptons, such as electrons, positrons, and their neutrinos. Soon, at higher temperature, more massive leptons – muons, antimuons, and their neutrinos – are generated and added to the dense lepton flood. From this era flee hordes of ghostly neutrinos that are with us to the present day. They are as numerous as the photons of the cosmic background radiation, and roam freely everywhere, passing unimpeded through the Earth and other celestial bodies. Neutrinos possess a nonzero but small mass, and being so numerous, they contribute much of the unseen mass of the present universe.

It is remarkable that with modern physics we can trace the history of the universe in broad outline back to the dawn of the lepton era, to a cosmic age of one ten-thousandth of a second, when the temperature is a few trillion degrees and the density nears a billion tons a thimbleful.

Before us lies the extreme early universe, the mysterious proterocosmos. Our overworked air conditioner is now at full blast and regrettably we can go no farther.

*　　*　　*

The dials on our timeship have registered shorter and shorter intervals of time, from billions of years to single years, then from years to seconds, then to milliseconds. Even though the steps taken have been progressively shorter, the cosmic scenery nonetheless has continually changed. The younger the universe, the faster it evolves, and in the very early stages everything changes with great rapidity. In front of us the proterocosmos looms as a blur of bewildering action. Perhaps as much of cosmic history lies ahead as behind us.

The complexity of the proterocosmos (the extreme early universe) reflects the complexity of the subatomic world and our understanding of what happens depends very much on the little we know of the world of subatomic particles.

Our timeship can go no farther. Standing at the dawn of the lepton era, sustained by little more than a spirit of speculative inquiry, we explore the proterocosmos by peering into the physicist's crystal ball.

*   *   *

Hadrons are the strongly interacting subatomic particles, such as protons, neutrons, and their antiparticles. Just before the onset of the lepton era, at a temperature of trillions of degrees and a density of billions of tons a thimbleful, the hadrons have their moment of glory as the dominant constituents of the universe.

Matter consists of particles (such as positive protons) and antimatter consists of antiparticles (such as negative antiprotons). The immense concentration of energy in the extreme early universe creates particles and antiparticles as fast as they annihilate each other. Matter coexists with antimatter.

Arthur Schuster in 1898 predicted the existence of antimatter in a letter to the science journal *Nature* entitled "Potential matter – A holiday dream". "When the year's work is over," he wrote, "and all sense of responsibility has left us, who has not occasionally set his fancy free to dream about the unknown, perhaps the unknowable." He went on to say:

> Surely something is wanting in our conception of the universe. We
> know positive and negative electricity, north and south
> magnetism, and why not some extra terrestrial matter related to
> terrestrial matter, as the source is to the sink.... Worlds may have
> formed of this stuff, with elements and compounds possessing

identical properties with our own, indistinguishable in fact from them until they are brought into each other's vicinity.

He wrote, "Astronomy, the oldest and most juvenile of the sciences, may still have some surprises in store. May antimatter be commended to its care!" and concluded with the words, "Do dreams ever come true?"

Paul Dirac in the late 1920s revived the possibility of antimatter in his work of uniting special relativity with the nascent theory of quantum mechanics. The positron (antiparticle of the electron) was discovered in 1932, and the antiproton in 1952; numerous other antiparticles have since been discovered.

Schuster guessed that antimatter has antigravity. In this he erred. Particles and their antiparticles exhibit opposite aspects, such as positive and negative electric charge, but both respond similarly to gravity. Schuster erred also in supposing that worlds of antimatter are as common as worlds of matter. Matter and antimatter, when brought together, annihilate each other with the release of considerable energy, mostly in the form of recognizable radiation. We do not see great quantities of this radiation betraying the presence of significant amounts of antimatter. The lost worlds of antimatter tell that our universe favors matter over antimatter.

A slight excess of matter over antimatter, of about one part in a billion, exists in the very early universe. Hadrons and their antiparticles are constantly created and annihilated. Annihilation, however, overtakes creation as the temperature drops. Matter and antimatter then disappear and only the slight preexisting excess of matter survives. This slight excess survives and now constitutes all the matter of the present universe. The immense energy released by the annihilation of the hadron hordes is eventually inherited by the cosmic background radiation.

We live in a topsy-turvy universe. The matter so vitally important in the making of galaxies is the result of a freakish and inconspicuous difference in the amounts of matter and antimatter in the proterocosmos. The cosmic radiation, enfeebled by expansion and seemingly of little consequence, is the legacy of the fantastic energy of the proterocosmos. What once was inconspicuous has become important, and what once was important has become inconspicuous.

*   *   *

Hadron particles are little worlds made cunningly. Composed of quarks, they dissolve into quarks at extremely high temperature. With the steady rise in density (we are still looking backward in time), the hadrons are squeezed together progressively more tightly, and finally they overlap one another. The hadron boundaries dissolve, the quarks burst free, and the universe transforms into a dense sea of free quarks. We have crossed into the bizarre quark era that stretches away to almost the beginning of time.

For fifty years, following James Clerk Maxwell's death in 1879, the universe was ruled by gravitational and electromagnetic forces. Sigmund Freud lamented, "With these forces nature rises up against us, majestic, cruel, and inexorable." But Freud was mistaken. Without gravitation there would be no galaxies, stars, and planets; without electromagnetic forces there would be no atoms, electricity, and light. Without either we could not exist.

We have nowadays the additional strong and weak forces. Without the strong force, protons and neutrons would not combine into atomic nuclei heavier than hydrogen; without the weak force, protons would not combine to form deuterium, and hence hydrogen would not burn to helium, and the stars would not shine over long periods of time. Also, without either we could not exist. The four forces of nature rise up not against us but for us, making life possible in the universe.

Albert Einstein, after his success at formulating general relativity theory, sought unsuccessfully to extend Maxwell's synthesis by combining gravity and electromagnetism within a unified geometric scheme. Sheldon Glashow, Abdus Salam, Steven Weinberg, and other physicists in more recent years, casting loose their moorings, have succeeded in showing that the electromagnetic and weak forces are dual aspects of a more fundamental electroweak force. At very high particle energies (or very high temperatures) these dual aspects fuse and become indistinguishable. The electroweak force in its unified form rules in the quark era. In the cool, dark universe of today it exhibits two guises – electromagnetic and weak.

Grand unified theories tie together the electroweak and strong forces into a hyperweak force. This force rules supreme before the onset of the quark era and disregards all distinctions between matter and antimatter, or between quarks and leptons. The universe consists of what might aptly be called *elem* (after ylem, introduced by George Gamow). At these ultrahigh energies, occurring when the universe is one trillion-trillion-trillionth of a second old, the hyperweak force merrily transforms matter into antimatter and back again, blithely switching quarks into antiquarks, leptons into quarks, and vice versa, rolling the strong, electromagnetic, and weak forces into one glorious free-for-all.

The sharp distinction between matter and antimatter comes with the decline of the hyperweak force and the rise of quarks and leptons at a time when the universe is only a trillion-trillion-trillionth of a second old. At about this time comes also the slight difference in the abundances of matter and antimatter. Because of the rapid expansion of the proterocosmos the various interactions and their decay schemes get out of step and matter is whimsically favored by one billionth more than antimatter.

Perhaps one day all four forces of nature will be happily combined into a supreme theoretical unity. Athwart the road to this desirable goal lies the sheer perversity of gravity. Herculean attempts by the cognoscenti to bring general relativity into the fold of quantum mechanics have so far not been very successful and we have yet to see whether grandiose schemes of supersymmetry, supergravity, and string theories can remove the roadblock.

*   *   *

The cosmos of long, long ago is popularly known as the "big bang." Fred Hoyle coined the vogue name "big bang" in 1950 in his sensational BBC lectures *The Nature of the Universe*. This catchy locution with its mythic overtones has misled many persons into the false belief that the universe originated as an explosion at a point in space.

At any moment the universe appears much the same everywhere in space. We cannot go to some other place in space and find the big bang waiting there to be discovered. We reach it by traveling not in a spaceship but in a timeship. (We travel not synchronically but diachronically.) The words "big bang" erroneously imply that the universe began as an explosion at a point in space. A deceived person might ask: Surely we can stay outside the big bang at a safe distance and watch the explosion? But the universe does not explode at a point in space. The big bang fills all space.

*   *   *

Quantum black holes – a basic constituent of spacetime – are the ultimate and most energetic of all particles. They have a mass ten billion billion times the mass of a proton (roughly the same as a speck of dust), and a size one ten-billion-billionth the size of a proton. The foamlike texture of spacetime consists of virtual quantum black holes, popping in and out of existence in unimaginable numbers, each savoring the joys of life for a Planck period. A Planck period equals one ten-million-trillion-trillion-trillionth of a second. These ultimate particles, forming the fabric of spacetime, have not been observed because of the extreme energy needed to lift them out of their virtual state. The likely existence of quantum black holes indicates that we can peer into our crystal ball back to a time when the universe has an age of one Planck period and a density of ten followed by ninety-three zeros times the density of water.

Our journey back through the proterocosmos comes to a final halt at an impenetrable barrier where the universe has an age equal to one Planck period. Around us lies a foam of indescribable chaos in which time and space are torn into discontinuities of cosmic magnitude. We can go no farther. An orderly historical sequence of events has ceased to exist, and past and future have lost meaning. Here, in the realm of quantum cosmology – the chaosmos – lie secrets that can foretell the design of the universe.

*   *   *

An attractive idea is that the universe begins in a state of utmost symmetry. But, like the symmetry of a pencil standing upright on its point, it is an unstable state. The evolution of the physical universe can be portrayed as a series of global transitions to successive states of progressively lower symmetry.

In the beginning the harmonious unity of all forces falls apart into two separate forces: the gravitational force and the hyperweak force. At first both are of equal strength. A remarkable and important transition to lower symmetry involves a phase change of elem, very much like the phase change of water to ice.

The physical universe begins at the Planck epoch ("Oh! let us never, never doubt what nobody is sure about!") and starts to expand. When the declining temperature reaches approximately ten billion billion billion degrees kelvin, the elem supercools and forms what Sidney Coleman called the "false vacuum." A property of the false vacuum is its negative pressure. Alan Guth showed in 1980 that as a consequence the universe expands exponentially while the density stays constant. He referred to this phase transition and period of accelerated expansion as inflation. Various opinions have been expressed concerning the duration of the inflation era but at present there is no consensus.

Negative pressure is just another name for tension. A stretched piece of elastic in a state of tension serves as an analogy of what happens in the inflation era. The stretching of the piece of elastic performs work and energy in the form of heat is generated. The greater the tension the greater the energy generated. Energy has mass. In a state of utmost tension the energy created maintains a constant mass density.

During inflation, which lasts plausibly only for a short time, the universe expands enormously. This has several important consequences. One is the dilution of monopoles. These ultrahigh-energy particles, a thousand times less massive than Planck particles, are initially profusely abundant. They are stable, which means they should nowadays still exist and be as abundant as the photons of the cosmic radiation. Inflation stretches the universe, the monopoles become widely separated and their contribution to the density of the universe is now vanishingly small. Inflation saves the universe from monopole domination.

Inflation also saves the day in another respect. Local tiny regions of the extreme early universe, so small they have had time to become homogenized, are stretched into vastly larger regions, each region retains its homogeneity, each is now much larger than the observable universe of today.

During inflation the density stays more or less constant but the temperature plunges. At the end of inflation the latent energy locked in the false vacuum breaks free and reheats the universe close to the original temperature at the beginning of inflation. The release of energy, caused by the loss of symmetry in the vacuum, populates the universe with a dense sea of quarks, leptons, and gluons. Before inflation the hyperweak force reigns, after inflation the electroweak and strong forces take over. It is thought that possibly the quantum fluctuations of the extreme early universe are stretched by inflation and become the density variation that evolve into galaxies.

* * *

"It's a poor sort of memory that only works backward," said the White Queen. While we still have the timeship we might as well try to explore the future. We leap forward in time into the far future. Again the dials record the passage of eons. After five billion years the Sun swells into a red giant and soon fades into a white dwarf. Terrestrial life has perished or fled elsewhere. After tens of billions of years guttering stars faintly illumine the fall of cosmic night.

Before us lie alternate cosmic scenarios. The universe either collapses and ends with the return of primeval chaos, or (at this moment the popular choice) expands endlessly in vacuous darkness. Either cremation on a flaming pyre or burial in a dark wall-less vault.

Consider the first alternative in which the universe ends in a blaze of light. After forty or so billion years expansion ceases and collapse commences. The galaxies begin to approach one another, and after a further forty or so billion years they arrive back where they are at present. Roughly ten billion years remain till the end of time.

The great dissolution begins. First, the clusters merge together, then the galaxies themselves overlap and dissolve. The universe now consists mainly of old stars (dwarfs, neutron stars, black holes) immersed in a commotion of gas. The big squeeze starts in earnest, and the ensuing tumult defies description. The poet Edgar Allan Poe more than a hundred years ago portrayed in *Eureka* the end of a collapsing universe. "Then, amid unfathomable abysses, will be glaring unimaginable suns...all this will be merely a climactic magnificence foreboding the great End." All astronomical systems are dismantled and all remnants of organic life obliterated. The stars accelerate and career about helter-skelter at higher and higher speeds, approaching the speed of light; a few collide headlong and erupt, but most wear away to nothing, leaving brilliant trails as they tear through the tumult. After a total lifespan of about hundred billion years the former brilliance returns and the universe reverts to primordial chaos. What happens then we do not know; perhaps a similar or an entirely different universe rises from the ashes of the old.

Consider the second and at present the more probable alternative in which the universe expands forever and ends not with a bang but a whimper. The visible stars and galaxies make up at most only ten percent of the energy content of the universe. The galaxies, their clusters, and all the matter they contain is like the tip of an iceberg. The remaining ninety percent exists in unknown forms. Perhaps it exists in the form of weakly interacting uncharged particles left over from the big bang, or perhaps it is a vestige of the immense energy latent in the vacuum. The latter possibility would explain recent observations indicating that the expansion of the universe is accelerating.

The galaxies, illumined by occasional flickers of light, voyage farther and farther apart in the dark abysm of time. The dials on our timeship whirl faster and after trillions of years all appears dead in an empty world of darkness.

Yet slow and unremitting agencies are at work. The galaxies and their clusters slowly shrink; many dark stars flee and take refuge on the outskirts of shrinking galaxies; many dark galaxies flee and take refuge in the widening abysses between the shrinking clusters. Also a steady outpouring of gravitational waves augments the slow collapse of stellar and galactic systems.

Hitherto we have merely dawdled on our journey. I set the timeship into overdrive and we leap forward across a trillion trillion trillion years and find that most matter is huddled into large black holes. On this vast time scale the hyperweak force has begun to exact its toll. Stars and galaxies that have escaped capture by black holes melt slowly away into radiation. Only black holes and weak radiation now exist. Across unspeakable spans of time we see the black holes themselves beginning to melt away.

Stephen Hawking of Cambridge University has shown that black holes emit radiation quantum mechanically. In the wavelike manner that protons penetrate each other's electrical repulsion barriers, particles tunnel through the surfaces of black holes and escape into the outside world. This occurs most easily at wavelengths roughly the size of the black hole. The emission of energy, mostly in the form of photons and neutrinos, causes black holes very slowly to lose mass.

A black hole of solar mass evaporates away in a time, measured in years, of 10 followed by 65 zeros. Black holes of galactic mass last much longer, and their lifespans are measured in googols of years. A googol – a term made popular by Edward Kasner and invented by his nine-year-old nephew – stands for the number 10 followed by 100 zeros. When our timeship is recording intervals of googols of years, the largest black holes have vanished. What remains is no more than the feeblest radiation forever becoming feebler.

Faster and faster we travel on a journey without end in a time machine whose controls have jammed and refuses to stop. Googol-plexes of years pass, where a googolplex is 10 followed by a googol of zeros. Googoogolplexes (my juvenile expression for 10 followed by a googolplex of zeros) of years pass in pitch-black emptiness, and still an eternity has just begun.

*   *   *

The simplest cosmological models show that a closed universe expands and then collapses, and an open universe endlessly expands. Observations by astronomers indicate that we live in a universe that endlessly expands. But for the last seventy years we have heard that the universe is open, and then, the next year that the universe is closed. We must sometimes doubt what everybody is sure about.

An eternal future offends against one's belief in a cosmos of rational unity. In a philosophical mood I like to think that a universe of rational unity must contain finite time as well as finite space, but most likely I am wrong. I share Emily Dickinson's modest prayer:

> And so upon this wise I prayed –
> Great Spirit give to me
> A heaven not so large as yours
> But large enough for me.

Only a cosmic jester would contrive a world of eternity and infinity.

*   *   *

Try to imagine a homogeneous universe having an infinite expanse of space. Out there, googols of light-years away, the cosmic scenery is much the same as here. Even if it is not the same, and the universe is inhomogeneous, and space is perhaps littered with an innumerable separate big bangs, every possible variation over googolplexes and googoogolplexes of light-years is endlessly repeated. On sufficiently vast scales an infinite universe is always homogeneous.

Our Solar System consists of a large but finite number of atoms. These atoms may be arranged into an enormous though finite number of different configurations. Each distinguishable configuration of this multi-solar-system ensemble has finite probability. Given an infinite number of solar systems, as there must be in an infinitely large universe, every configuration of finite probability is repeated an infinite number of times. Nothing is unique. Out there exists an infinite number of identical Solar Systems, with identical Earths, with identical human populations. Each of us at this moment is doing the same thing in an infinite number of places. What can be the point of all this multiplicity when once is often more than enough?

*   *   *

We have a fascinating view of the physical universe that may be shaken into many kaleidoscopic patterns. We marvel at these patterns, setting aside most as implausible, selecting a few, usually the simplest, consistent with what we observe and know. We think the universe to be at least as complicated as anything it contains. A single subatomic particle is still a bewildering mystery; can the physical universe that supposedly makes sense of it all be any simpler?

Part III   The Cloud of Unknowing

14 The Witch Universe

Francis Bacon, English courtier and statesman of the late sixteenth and early seventeenth centuries, promoted a philosophy of empirical science and declared, "let every student of nature take this as his rule: that whatever the mind seizes upon with particular satisfaction is to be held in suspicion." His strong belief in empirical methods of inquiry assured him that witches existed.

Let us look at the witch universe in which this incredulous and illustrious man lived.

*   *   *

Tracing the development of ideas in the long Middle Ages leads the student into a bewildering labyrinth of astonishing beliefs. The works of Jabir ibn Haiyan, court physician in the eighth century to Harun al-Rashid (the caliph of Baghdad famed in *The Thousand and One Nights*), became widely known for their medical lore and learned alchemy. Jabir was later latinized into Geber, and because of the rigmarole and obfuscation of the numerous works attributed to him, the word Geberish became eventually gibberish.

In the Middle Ages the telluric elements of earth, water, air, and fire exhibited respectively the qualities of cold, wet, dry, and hot. By erudite argumentation the elements accounted for bodily humors of melancholy, phlegm, choler, and blood, which in a marvelous manner corresponded with the characteristics of creation, fall, redemption, and judgment. Acts of the will were governed by God, acts of the intellect by angels, and acts of the body by the celestial orbs. Each person possessed a daemon or genius who acted as a guiding spirit. Less remarkably, the ten wits comprised the five outer senses of sight, hearing, smell, taste, and touch, and the five inner senses of memory, thought, imagination, instinct, and common sense.

Metals according to alchemists and astrologers possessed intimate relations with the celestial bodies: silver with the Moon, quicksilver with Mercury, copper with Venus, gold with the Sun, iron with Mars, tin with Jupiter, and lead with Saturn. The laboratorium of the medieval alchemist was lavishly equipped with vessels, vials, urinals, alembics, descensories, sublimatories, and indeterminate paraphernalia, with furnaces and various contraptions installed for research in calcination, sublimation, distillation, condensation, and solification. Chaucer in *The Canon's Yeoman's Tale* gives us a peep into the medieval laboratory, and we see how richly it is stocked with alkali, arsenic, brimstone, sal ammoniac, saltpeter, quicksilver, vitriol, and herbs of numerous kinds. The abracadabra, incantations, obscene prescriptions, and the rest seem to us pure gibberish.

The consuming passion of the alchemists, not unlike our modern quest for the cure of cancer, was the search for the philosopher's stone that when discovered would transmute base metals into gold and restore lost youth.

* * *

Kings believed that by augmenting their power they implemented the divine intention. As their authority grew in the late Middle Ages, the power of the feudal nobility waned and European nations drifted apart.

Wider horizons, strange foreign lands, new lifestyles, and novel schemes of thought demanded a more capacious world view, and the medieval universe was soon bursting at the seams. Its decline in the late Middle Ages and fall in the Renaissance was accompanied in Western Europe by widespread social unrest and upheaval. Warfare punctuated by plagues erupted and became a way of life, arresting and even reversing the population growth.

The Church ceased to have a civilizing influence and seemed bent on wrecking all it had accomplished. The assertion that Jesus was a poor man affronted rich prelates and was condemned as heresy by popes. Corrupted by power, riddled with simony, demoralized by the crusades, and lost in the casuistry that glorious ends fully justify inglorious means, the Church was incapable of implementing long-overdue reforms. In vain, Erasmus and other enlightened humanists endeavored to moderate rising social tensions with counsels of forbearance. Reformation and Counter Reformation armies marched and countermarched amidst the ruins of the medieval universe.

Arts, classics, drama, and poetry became havens in which distraught sections of the public sought shelter from a grim and cruel reality. Painters in brilliant colors glorified the human figure against mythical backgrounds; poets in wondrous words rhapsodized on themes of passionate love; dramatists in marvelous fantasies catered to enthralled audiences. Thousands of plays opened up escapist worlds, and one-tenth of the adult population in London could be found of an afternoon at the theaters.

It was an age of disillusion, of make-belief, of paradise lost in a ravaged world. We call it the Renaissance and forget that the engines of art are fueled by the distillates of anguish.

Out of the despair of a world in disarray came a pathological desire to find scapegoats. First the Jews, who denied the godhead of Christ, had poisoned the wells and caused the plagues. Then the witches with their black arts had formed a conspiracy with the Devil and caused all the misfortunes of the sixteenth and seventeenth centuries. The Church, backed by the secular authorities, led the campaign of persecution.

The crusades had shown that wars waged against external enemies were an effective way of distracting the public. After the collapse of the Christian venture in the Holy Land, the grip of central authority was maintained and tightened by stirring up fears of internal enemies.

* * *

Witchcraft until the thirteenth century consisted mostly of superstitions and folk beliefs that added variety to the lives of both highborn and lowborn, and much the same as today accounted for the fortunes and misfortunes of life when other explanations seemed unconvincing. The arts and crafts of witchery covered a broad spectrum, ranging from the benign to the malign; from healing, dispensing herbal preparations, midwifery, charms, spells, and love potions, to casting the evil eye, mutilating waxen images, invoking the dead, contriving miscarriages, frigidity, and impotency, conjuring up storms and floods, causing diseases, and even death. To cultured clerics of the day, witchcraft was deplorable, and much of it seemed ludicrous and unrelated to Christianity.

The witch hunt began in the late Middle Ages and developed into the witch craze of the Renaissance. By the sixteenth century witchcraft had become a reality of supreme significance, and all lingering doubts about its diabolical nature had vanished. Even folk dances and festivals of pre-Christian origin were associated with the machinations of witchcraft and devilry. Biblical texts on the subject of witches and demons (with the biblical command "thou shalt not suffer a witch to live") supplied the principles of a new world view.

The witch universe was a dark, inverted image of Christianity. As seen in this demented universe, a hideous conspiracy with the Devil had mantled the Earth in a sinister twilight. The Lords of Light and Darkness were at last face to face, locked in a life-and-death struggle. In the desperate war of the worlds that ensued, involving all sections of the public and mobilizing all resources of church and state, fires were stoked, torture chambers equipped with the latest weapons, and monkish armies recruited to fight against an invasion of horrifying demons. The blasphemous, sacrilegious, cannibalistic witches were devil-worshipping demons; they flew through the air on besoms or grisly beasts to clandestine covens, assumed grotesque animal shapes, killed and ate young children, desecrated the Cross, and were associated in the minds of celibate priests with the most obscene sexual practices. Such was the calamitous state of European society after the collapse of the medieval universe.

*   *   *

In the high Middle Ages, in 1258 and 1260, Pope Alexander IV issued bulls bidding the Franciscans and the Dominicans to refrain from judging witchcraft unless heresy was amply demonstrated. Two centuries later, in the Renaissance, the outlook had totally changed: the arts of witchcraft and the most heinous forms of heresy had become one and the same.

The fables of country folk and townsfolk were little more than vestiges of old pagan beliefs about hobgoblins, fairies, and the like. Zealous clerics in the late Middle Ages, increasingly fearful of nonconformity in a society of growing complexity, exaggerated these popular credulities, and with fevered imagination invested them with elaborate sophistication.

Whenever the inquisitors questioned the members of a dissenting sect or dissident cult – such as the Cathars and the Waldensians – they found that all dissenters and dissidents had similar revolting vices: they killed babies by tossing them from one person to another, then ate their bodies, had identical vicious and horrid habits, held sexual orgies at nocturnal meetings, and called up the Devil, who appeared as either a lascivious man or a noisome beast. Simple and uneducated persons were found guilty of abominable practices that scholars trawled from mythology and dredged up from the history of olden times that the population previously had not the slightest knowledge. Peasants, distinguished persons, and even the proud Knights Templars were the innocent victims of a demonizing mania that gathered momentum and swept through Europe. The method of interrogation, legitimized by the popes, consisted of questioning the accused victim about his heretical and gruesome habits that were described to him in lurid detail. When the accused denied such behavior, torture succeeded sooner or later in extracting the required admission. Torture could be mitigated by confessing the names of other heretics. In this way the fantasies rife in the overcharged minds of the inquisitors became established truths, and society accepted the reality of a witch universe foisted on it by self-deluded intellectuals.

*　　*　　*

The witch craze began with the bull of Pope Innocent VIII issued in 1484 that deplored the widespread depravities of witches. In bitter sorrow the pope enumerated the appalling enormities of witchcraft:

> Many persons, of both sexes, unmindful of their own salvation and straying from the Catholic Faith, have abandoned themselves to devils, incubi and succubi, and by their incantations, spells, conjurations, and other accursed charms and crafts, enormities and horrid offenses, have slain infants yet in the mother's womb...these wretches furthermore afflict and torment men and women, beasts of burthen, herd-beasts, as well as animals of other kinds, with terrible and piteous pains and sore diseases, both internal and external; they hinder men from performing the sexual act and women from conceiving...and at the instigation of the Enemy of Mankind they do not shrink from committing and perpetrating the foulest abominations and filthiest excesses to the deadly peril of their own souls, whereby they outrage the Divine Majesty and are a cause of scandal and danger to very many.

In this bull the pope appointed "our dear sons Heinrich Kramer and James Sprenger, professors of theology, ... as Inquisitors of these heretical pravities," to go forth into northern Germany and wherever else they could find "the disease of heresy and other turpitudes diffusing their poison to the destruction of many innocent souls." The pope empowered them to condemn and punish the offenders.

Kramer and Sprenger were distinguished on account of their vigilant defense of Christian society, inquisitorial zeal, and diligent burning of heretics. They sallied forth at the pope's command and three years later reported the results of their investigations. Their report was published under the title *Malleus Maleficarum* or *The Hammer of Witches* (the full title reads, *The Hammer of Witches that Destroyeth Witches and their Heresy as with a Two-edged Sword*). Some details of this report deserve quoting, for in restrained and moderate language (by the standards of those times) the cosmologists Kramer and Sprenger reveal to us a picture of the witch universe in the making.

Three witches burning, Germany 1555.

They find, reported in the *Malleus Maleficarum*, that towns and countryside swarm with astrologers and sorcerers consulted by all classes, low and high, and that political plottings and royal schemings are everywhere associated with the black arts; they notice "in this twilight and evening of the world, when sin is flourishing on every side and in every place, when charity grows cold and the evil of witches and their iniquities superabounds," how evident it has become that "witches and the Devil always work together, and that insofar as these matters are concerned, the one can do nothing without the aid and assistance of the other;" and for the benefit of the reader they outline the various branches of witchcraft, present numerous anecdotal accounts of witchcraft, and specify in detail the infamies and evils practiced by witches; thus of the forty-one witches burned in 1485 at Como, they affirm that these creatures had intercourse with the Devil, and "this matter is fully substantiated by eye-witnesses, by hearsay, and the testimony of credible witnesses;" for the witches have lewd practices in order that they may increase to the detriment of the Faith, and in their homes are visited by the Devil who copulates with the he-witches as a succubus and the she-witches as an incubus, and the semen taken from one and given to the other breeds more and worse witches; women in particular are vulnerable to the wiles of the Devil, for they are feeble in mind, credulous, more carnal than men, and liars by nature, and the works of the blessed saints and holy fathers attest that "all witchcraft comes from carnal lust, which is in women insatiable" (for "what else is woman but a foe to friendship, an inescapable punishment, a necessary evil, a natural temptation, a desirable calamity, a domestic danger, a delectable detriment, an evil of nature, painted with fair colours!"); and note that one of the aims of witches is to kill as many children as possible before Baptism, thus debarring them from Heaven and preventing the Elect from reaching the final number, thereby delaying the Day of Judgment in this twilight age; witches eat children, and from the bodies of children – often stolen from graves – they concoct obscene unguents essential in their transportations and transformations; witches stir up hailstorms, raise tempests, and cause thunderbolts to blast men and beasts, and are an abomination and the worst of heretics; those bishops and all rulers who fail to essay their utmost in stamping out witchcraft must be judged as abettors and punished as heretics; "any witness may come forward" and the "accused person, whatever his rank or position, upon such an accusation may be put to the torture, and he who is found guilty, let him be racked, let him suffer all other tortures prescribed by law in order he be punished in proportion to his offenses;" and nowadays instead of being thrown to wild beasts as in olden days, "they are burnt at the stake and probably this is because the majority of them are women;" if the accused demands to confront her accusers, or have a legal advocate, it is up to the judge to decide, and he must take into account that such matters always delay and confuse the proceedings, and must weigh the possibility that the accused speaks with the voice of the Devil; each witch must be stripped ("by honest women of good reputation"), shaven of all hair, and put to the question while naked; at first, in justice, she must be questioned lightly without shedding blood, even though moderation in questioning is always fallacious and generally ineffective; she must be questioned about her intercourse with the Devil, the infants killed and eaten, and other matters of which she is justly accused; note that witches frequently commit suicide between the periods of examination and must be prevented by shackling, for they are induced to do so by the Devil; whenever witches confess, they are prompted always by a divine impulse from an Angel, but when they plead innocence or do not confess or have "the evil gift of silence that is the bane of judges" they are so impelled by the Devil, and the interrogation must continue; witches who cry out under investigation give voice to evil impulses, and yet those who cannot weep are manifestly possessed by devils; weeping may itself be an artifice of the Devil seeking to avert justice by creating pity; those who expire during the examination are saved by the Devil for the evil purpose of escaping full confession; a judge or inquisitor may facilitate his proceedings by promising the accused her life if she fully confess and name her evil heretical collaborators, although under no circumstances must such promises be kept, and may be made with "the mental reservation he will be merciful to himself or the state," for "whatever is done for the safety of the state is merciful;" when a confession has been extracted, the witch must then be handed over to the secular authority for execution, because the Church cannot exact the ultimate penalty; divine providence has arranged that all inquisitors and judges are immune to the powers of witches, and they should therefore never be deterred in their proceedings; and inquisitors and judges must understand that witches are the incarnations of Evil who have made a compact with the Devil and are an affront to the Terrible Judge, and must remember that witches, "however much they are penitent and return to the faith, must not be punished like other heretics with lifelong imprisonment, but must suffer the extreme penalty," and inquisitors and judges must bear in mind that if those "who counterfeit money are summarily put to death, how much more must they who counterfeit the Faith."

While numerous supplementary texts described, explained, and filled in the details, the *Witch Hammer* became the standard textbook, the *Principia* of a new world. For more than two centuries it had enormous influence and was used by Catholics and Protestants alike. In the hands of every scholar and on the desk of every official it dispelled all doubt concerning the evil reality of witches; it spurred the fainthearted and urged to even greater efforts the zealous in their constant battle against witchcraft.

A devout Christian had no choice other than to support fully and openly the eradication of the dark forces subverting society; had no choice other than to bear witness whenever necessary against witchcraft tendencies in spouse, parent, brother, sister, son, daughter, or other relative, and neighbors.

*   *   *

In the Renaissance, with its "rebirth in the nobility of the human spirit," every scholar and peasant shared the same fixed beliefs concerning witches and knew they flew through the air to their sabbats held in caves, on mountaintops, and other eerie places, where they gathered at cannibalistic feasts, cavorted to macabre music, indulged in sexual orgies, parodied Christian ritual, renounced God, and worshipped the Evil One who appeared before them in various dreadful guises.

And the awful fact was that wherever you found one witch and used the just and legitimate instruments of inquiry, you inevitably found many others. Their numbers multiplied and seemed without limit. Male and female witches and their evilly spawned children were consumed by fire in mounting numbers, and still they multiplied. Trevor-Roper in *The European Witch Craze* depicts the scene:

> All Christendom, it seems, is at the mercy of these horrifying creatures. Countries in which they had previously been unknown are now suddenly found to be swarming with them, and the closer we look, the more of them we find. All contemporary observers agree that they are multiplying at an incredible rate. They have acquired powers hitherto unknown, a complex international organization, and social habits of indecent sophistication. Some of the most powerful minds of the time turn from human sciences to explore this newly discovered continent, this America of the spiritual world.

The details they discover are amply confirmed by experimentalists working in the confessional and torture chamber, by theorists working in the library and cloister, leaving the facts more secure and the prospect more alarming. Instead of being stamped out the witches increase at a frightening rate, until the whole of Christendom seemed about to be overwhelmed by the marshaled forces of triumphant evil. To protest against witch hunting as inhumane in a time of dire emergency was unthinkable, condemned by the popes as bewitchment and the result of consorting with devils.

Who were the witches? Apparently they were the old and lonely, mostly female, the ugly and crippled, the weak and sick in mind, the insane, the hated who could be spited by false testimony, the vulnerable whose property could be confiscated, strangers, dissenters, as well as charlatans, poisoners, and other malefactors. Protests of innocence had no effect other than to confirm the alleged witchcraft. Confession by the accused and the naming of confederates offered the only escape from torture, followed by burning at the stake.

It was an age that fully believed in witchcraft, and the accused shared the prevailing beliefs. Many of the accused suffered from hallucinations and thought they possessed the alleged witchcraft powers. Incarcerated under dehumanizing conditions, fed with an enfeebling diet, then put to the question, almost all were induced to think they were guilty of the alleged offenses. The number of victims burnt at the stake is unknown; various estimates place the total somewhere between a few hundred thousand and more than a million.

Not the humanities, not religion, but the sciences saved Europe from the mad witch universe of the Renaissance. While contributing to the demolition of the medieval universe, the sciences were reaching out to a world view more capable of defining the limits of human control over nature, in which in later years the gibberish of demonology seemed incredible.

The emergence of science, says Herbert Butterfield in *The Origins of Science*, "outshines everything since the rise of Christianity and reduces the Renaissance and Reformation to the rank of mere episodes," and "looms so large as the real origin of the modern world and its mentality that our customary periodisation of European history has become an anachronism and an encumbrance."

*　　*　　*

No one knows exactly what constitutes the "scientific method." Francis Bacon, a persecutor of witches, declared that science is wholly empirical, hence fully inductive (all conclusions are drawn solely from observations). Empiricism is the philosophy that the scientific method consists of performing observations with an open mind and drawing common sense conclusions. It is a truth-seeking method of inquiry that fully confirmed Bacon's belief in the reality of witchcraft. Seeing is believing, which Bacon emphasized, but also believing is seeing, which he failed to acknowledge. The scientific method, like any method of inquiry, requires the exercise of imagination before, during, and after an observation.

The witch universe was vividly real in the minds of those who lived in the Renaissance. It was believed to be true by inquisitors, jurists, victims, and every section of the public. It had its graduate textbook *Malleus Maleficarum*; its facts were investigated and repeatedly verified with the instruments of the torture chamber. The phenomena of demonology and the reality of witchcraft were repeatedly tested and verified. To say these instruments of inquiry are not those used in a modern laboratory misses the point. In the context of the belief system of that time they were truth-seeking and fully appropriate. The facts stood out stark and clear, plain for all to see, unalterable and fully verified. "Would to God," said Kramer and Sprenger, "we might suppose that all this to be untrue and merely imaginary, if only our Holy Mother the Church were free from the leprosy of such abominations." Were we living in the Renaissance, we probably would have thought the same.

Science explores, defines, and explains the world around, and known facts are repeatedly tested and verified. The inquisitors also explored, defined, and explained the witch universe and repeatedly tested and verified their facts. In general, it seems that each universe is self-affirming and determines its own rules of verification.

An hypothesis may be verified many times by observation and accepted as the truth until contradictory facts emerge. Thus, all swans are white is verified by repeated observations and accepted as the truth until reports reach us of black swans in Australia. Empirical knowledge is verified by observation until falsified by contradiction. The philosopher Karl Popper has argued that science consists of facts and theories that are vulnerable to falsification. But what is falsifiable in any age depends very much on the world view and prevailing beliefs and its criteria of what constitutes valid knowledge. In a scientific age what is falsifiable distinguishes science from nonscience, and science determines what is falsifiable. The physical universe is falsifiable according to the principles of that universe. The witch universe was also falsifiable according to its principles. All universes are falsifiable within their own terms of reference.

In the physical universe we regard the content of the physical sciences as falsifiable, and we use falsification as the demarcation between acceptable and unacceptable scientific knowledge. In the mythic universe we regard the gods and their works as falsifiable and we use falsification as the demarcation between acceptable (devout) and unacceptable (heretical) knowledge. Vulnerability to falsification distinguishes between things that fit naturally and things that fit unnaturally in the universe we happen to occupy. In any universe what is fitting is falsifiable and what is unfitting is either falsified or unfalsifiable. In the mythic universe we have the testimony of the prophets and saints that angels exist and propel the planets. In the modern physical universe the myth that angels propel the planets is falsified, and the myth that angels exist is unfalsifiable.

15 The Spear of Archytas

The Universe is everything. It includes us and the rational universes we collectively devise. Each universe unifies a society and dictates the "true" facts. Individuals suppose with unfailing confidence that their particular universe is the Universe, and their confidence is not in the least shaken by the fact that our ancestors lived in very different universes and our descendants in the future will live also in totally different universes. In all universes things have their causes, often hidden from ordinary mortals. We depend on our wise men – emperors, kings, shamans, priests, sages, prophets, and professors – to put us right and tell us the "true" facts. As long as somebody reliable knows the truth, all is right with the universe.

*   *   *

All universes have their rules of containment that define what is included as fitting and what is excluded as unfitting. Thales said the Ionian universe consisted of water; Anaximenes said air; Heraclitus said fire; Xenophanes said earth; Empedocles said earth, water, air, and fire. Democritus said the Atomist universe consisted only of atoms and the void; all else was illusion and opinion. Plato said the Platonic universe consisted of the eternal verities of the Mind; all else was shadow and deception. Aristotle said the Aristotelian universe consisted of earth, water, air, fire, and ether in ascending order, animated by Ideas, and nothing existed beyond the sphere of the stars. Saint Augustine said the Christian universe consisted of the Word of God, and all else was heresy. John Wheeler of Princeton University said the physical universe consisted of "empty curved space," and "matter, charge, electromagnetic and other fields were manifestations of the curvature of space."

In every age people believe that their universe contains all that is believable and real. Wise men in their palaces, temples, academies, and universities reject the rest as opinion and illusion. Forget all the superstitions of the uneducated and the myths your parents taught you. For behold! Here is the true universe, awesome, vast, and wondrous. The world is an immense tug-of-war with gods and demons pulling on a giant serpent; the world is the handiwork of almighty gods whom we must obey and worship or reap the misfortune of their wrath; the world is a finite geocentric unity of crystalline spheres; the world is a dance of atoms and waves, all else is outworn myth and discredited theory. The scene is timeless. Yesterday there is a false image, today the true face.

In *Our Mutual Friend* Charles Dickens wrote, "Mr. Podsnap settled that whatever he put behind him he put out of existence... he had even acquired a peculiar flourish of his right arm in often clearing the world of its most difficult problems by sweeping them behind him." The Podsnap flourish is encountered in all walks of life and whatever fails to fit is swept out of existence. What is not contained does not exist. Many persons, including scientists, believe that what is not contained in the modern physical universe does not exist and is waved aside with the Podsnap flourish.

Rules of containment protect the universe from being swamped by the errant fancies of individuals. Without such rules defining what is rational and fitting, rejecting what is irrational and unfitting, the universe would collapse and society break down.

*   *   *

The physical universe contains all that is physical and nothing else. This is the containment principle of the physical universe. Atoms and cells, flowers and planets, stars and galaxies are physical things as studied by the natural sciences. Atoms and their electron waves, reproducing organisms and their differential survival, evolving stars and their transmutation of elements, space and time and their dynamic warping, all belong to the physical universe that contains physical things and nothing else. DNA molecules with their genetic coding, our bodies born and doomed to die, our brains throbbing with bioelectronic activity are the stuff of the physical universe. But the mind and its consciousness are unphysical, hence uncontained and therefore "unreal" except for analogous forms of cerebral activity and glandular chemistry.

Learned savants assure us that our modern universe is the Universe, that all happenings have their physical causes, all objects consist of molecules, atoms, and subatomic particles, and what is unphysical does not exist in the real world. They teach us that organic life evolved over billions of years on Earth, that our spinning Earth orbits the shining Sun, that our Sun is a star far from the center of the Galaxy, that our whirligig Galaxy of stars is but one in a vast concourse of galaxies, that the universe originated as a big bang and is still expanding.

The physical universe contains all that is physical and nothing else. This is an eminently sensible principle provided we do not inquire too closely into the meaning of "physical." Dr. Johnson's stone has lost its concreteness in a cloud of probability waves. The material universe in which everything was a game of billiards has become a universe of virtual waves that collapse into definite particle states when we make observations.

The first and most important containment rule of the modern world is that space and time are physically real – real as apple pie – and spacetime is a curved dynamic universal constituent of the universe. Space and time are intimately wedded and contained and therefore cannot extend beyond the universe. This has important consequences.

The second rule, more difficult to grasp, can be summed up by saying that images do not contain the image maker. A universe does not contain the person thinking about that universe. A universe is a product of the mind and contains, at best, a representation of that person's mind. Much credit for this remarkable discovery goes to the modern physical universe and its clear concepts of what is physical.

The cosmic edge and the spear of Archytas. (E. Harrison, *Cosmology: The Science of the Universe*, 2nd edition, Cambridge University Press, 2000.)

Containment causes much perplexity. In times of stress we seek shelter in the comfort of myths that at other times we deny as superstition. We agonize over puzzles such as free will and the problem of how nerve impulses translate into thoughts. Our anguish is needless. Freewill and the mind with its thoughts belong to the Universe; determinism and the brain with its nerve impulses belong to the universe. The recognition that the uncontained may actually exist in the Universe, though not in a universe, cuts like a bracing wind through a history of discourse.

*   *   *

General relativity has shown that space and time are physically real and are therefore contained in the physical universe. By journeying in space and time we are unable to escape from the universe. The ancient view, popular until recent times, held that space and time contain the universe. The modern view is the opposite: the universe contains space and time.

Cosmic edges defining the limits of the universe were once of enormous importance in cosmology. A cosmic edge implied a cosmic center. The center was first the nation, then the Earth, the Sun, and finally the Galaxy. The edge was less obvious and posed problems that taxed the ablest minds.

What happens when a spear is thrown across the cosmic edge? This is the celebrated cosmic-edge riddle popular in the Middle Ages. It can be traced back to Archytas of Tarentum, the Pythagorean philosopher–scientist, statesman, soldier, friend of Plato, and possibly the model for Plato's philosopher–king. The spear of Archytas is the shatterer of universes. The Roman poet Lucretius used the riddle of Archytas to great effect:

> Suppose for a moment that the whole of space were bounded and that someone made his way to the outermost boundary and threw a flying dart. Do you choose to suppose that the missile, hurled with might and main, would speed along the course on which it was aimed? ... With this argument I will pursue you. Wherever you may place the ultimate limit of things, I will ask you, "Well, then, what does happen to the flying dart?"

Does the spear rebound, continue on its way, or disappear? Lucretius gave the Atomist answer: "Learn, therefore, that the universe is not bounded in any direction."

Simplicius in the sixth century (early Middle Ages) referred to Archytas in his commentary on Aristotle's *Physics*: "If I am at the extremity of the heaven of the fixed stars, can I stretch outwards my hand or staff? It is absurd to suppose that I could not; and if I can, what is outside must be either body or space. We may then in the same way get to the outside of that again, and so on; and if there is always a new place to which the staff may be held out, this clearly involves extension without limit." This fragment remained untranslated into Latin until the sixteenth century.

Lucretius's epic poem *The Nature of the Universe* influenced Cardinal Nicholas of Cusa who saw in the infinity and ubiquity of God the justification for a centerless and edgeless universe. Possibly the poem influenced the astronomer Thomas Digges, who tore away the outer boundary of the Copernican system, and the physician William Gilbert who advocated the infinite Atomist universe of numberless celestial worlds, and the luckless Giordano Bruno who was burned at the stake for holding such views. In his *Infinite Universe*, Bruno wrote, "If a person would stretch out his hand beyond the convex sphere of heaven, the hand would occupy no position in space, nor any space, and in consequence would not exist.... Thus, let the surface be what it will, I must always put the question: what is beyond?" Only one answer resolved the riddle: the universe is edgeless, hence limitless in extent.

In the Cartesian and Newtonian systems of the world space was edgeless. John Locke in *An Essay Concerning Human Understanding* (1690), expressed the new view: "For I would fain meet with that thinking man that can, in his thoughts, set any bounds to space; more than he can duration; or by thinking, hope to arrive at the end of either." The spear of Archytas had demonstrated that space had no edge. Geometers supposed that edgeless space necessarily extended in all directions to infinity. Curved space (like the two-dimensional surface of a sphere), introduced in the middle of the nineteenth century, made possible the idea of edgeless space that is finite in extent.

We can identify wall-like, clifflike, and marshlike edges. First and most ancient was the view that the universe ended abruptly at a wall-like edge, as in a giant egg or cave. Lucretius, the Epicurean, echoing the Atomists, rejected this answer, arguing that space is continuous and cannot end: "It is a matter of observation that one thing is limited by another. The hills are demarcated by air, and air by hills. Land sets bounds to seas, and seas to every land. But the universe has nothing outside to limit it." Johannes Kepler, who abhorred the notion of a boundless universe, was not convinced. At night, said Kepler, when we look out between the stars we see an enclosing dark wall. Imagine what might befall us if the universe stretched away endlessly, populated with numberless stars, as had been claimed by the ancients. Every line of sight would terminate at the surface of a distant star and every point of the night sky would blaze with starlight. The sky everywhere would be as bright as the Sun.

Why is the night sky dark in a boundless universe? This riddle, deriving indirectly from Archytas's cosmic-edge riddle, remained unsolved until recent times. Light has finite speed and when we look out to vast distances we also look far back to a time before the birth of the first stars. We cannot look out to unlimited distances in a universe of finite age, and the total number of observed stars is insufficient to cover the whole sky. Whether the universe expands or is static, whether it consists of stars distributed uniformly or clustered into galaxies, and whether it is open (spatially infinite) or closed (spatially finite), we see dark gaps between the stars. Thus, the sky is not ablaze at every point with bright starlight.

The second was the clifflike edge adopted by the Stoics of the Greco-Roman world. In the Stoic universe the stars were distributed within a finite spherical cosmos, and beyond the edge of the cosmos extended infinite empty space called the Void. The spear crossed the edge of the cosmos and was lost in the Void. The Void was added to the Aristotelian system in response to the riddle of Archytas. William Herschel adopted the Stoic picture in the eighteenth century, and in the early decades of the twentieth century the idea of an island universe – our Galaxy surrounded by an endless void – gained wide support. Harlow Shapley, a famous astronomer early in the twentieth century, tried to resolve the dark-sky riddle by adopting a Stoic model of the universe. Astronomers have since discovered that the universe of galaxies extends to enormous distances with no evidence of a clifflike edge.

The third was the marshlike Aristotelian edge. In the Aristotelian world view, material bodies existed only in the sublunar sphere. If the spear passed beyond the lunar sphere, it became ethereal and its natural and only possible motion was circular around the Earth. The physical realm merged into the etheric realm and no sharp boundary between the two existed. The thrust of "with this argument I will pursue you" was lost and the pursuer led into a metaphysical marshland where the argument lacked cogency. The Aristotelian edge lingered on. I still remember the teacher in a scripture lesson pointing to the ceiling when asked where is God, and saying, "Up there in Heaven." A decade earlier, Einstein had shown with the theory of general relativity that space and time are physically real. More than two thousand years previously Anaxagoras had said the laws and constituents of the universe are the same everywhere. A nonphysical realm does not exist in the space and time of the physical universe.

Beginners in modern cosmology often have in mind a picture resembling the Stoic cosmos. They suppose that space extends into a void from which it is possible to view the expanding universe. From this extracosmic grandstand they see the universe as an exploding cloud consisting of a swarm of galaxies and stars with the big bang at the center. This picture, perpetuated in popular literature, is misleading because no outside void exists from which the universe can be observed. The big bang did not occur at a point in space, for it occupied the whole of space.

Cosmic edges have gone and with them the cosmic center. Physical space is continuous and is either infinite or finite in extent.

*　　*　　*

A miscellany of subjects falls under the heading of containment. Some are elementary. Obviously, contained things are neither larger nor older than the universe; a galaxy or supercluster of galaxies is necessarily smaller and younger than the universe.

Creation is a particularly interesting subject. In broad terms, creation has three distinct meanings. Scientists apply the word creation to a physical change in state in which fundamental quantities remain conserved. Thus, the creation of a particle and its antiparticle conserves energy and electric charge. Also, the creation of structure in the universe conserves basic physical quantities. The origin of life on Earth is a physical form of creation.

Then we have the metaphysical (or miraculous) kind of creation in which something is created out of nowhere at a place in space at a moment in time where previously there was nothing or something quite different. This kind is common in mythology.

Lastly, we have cosmogenesis (the creation of the universe), in which "God created the heaven and earth."

Creation is either physical, metaphysical, or cosmogenic. The first two deal with creation in the universe, and the third deals with creation of the universe. We have two basic kinds: the physical and metaphysical kind refer to contained creation, the cosmogenic kind refers to uncontained creation. The first deals with creation in space and time, and the second deals with creation of the universe that includes space and time.

Hermann Bondi, Thomas Gold, and Fred Hoyle in 1948 proposed a steady-state expanding universe in which matter is continuously created everywhere in space. They discarded conservation of matter and replaced it with what seemed to them more fundamental: the conservation of the universe in its present state. In an expanding universe the widening gulfs of space between old galaxies become occupied by new galaxies born from newly created matter, thus preserving the appearance of the universe. In the controversy following this proposal many contestants held the view that the instant creation of a big-bang universe is of the same kind as the continuous creation of a steady-state universe. We have the choice, it was said, of creation of the universe all at once or little by little. But the creation of a whole universe is very different from the creation of its bits and pieces; the former is uncontained, the latter contained.

Both physical and metaphysical creation occur in the space and time of an existing universe; cosmogenesis is the creation of the whole universe including its space and time. Failure to distinguish between contained and uncontained creation violates the rules of containment.

Cosmogenesis cannot have the same meaning as miracle-making unless we revert to the view that the universe is contained in a preexisting absolute space and time. Nowadays, we are not free to say that at a place in space there is nothing and an instant later at the same place there is a universe.

Some cosmologists in the past have strongly disliked the idea of cosmic birth and death. Most notable was Arthur Eddington, who in 1930 constructed a model universe having an infinite past and an infinite future, thus avoiding a cosmic beginning and end. With his eternal universe he sought to escape the mythological specter of cosmic creation and annihilation. But he failed because cosmogenesis is the creation of a whole universe containing time. The physical universe is not created in time, and cosmogenesis cannot in principle be pushed out of sight into an infinite past. A universe having an infinite span of time is created just the same as a universe with only a finite span. From a cosmogenic viewpoint the eternal Eddingtonian and steady-state universes are the same as any other. All universes of finite or infinite duration in time, of finite or infinite extension in space, with or without big bangs are confronted with the problem of cosmogenesis.

We must realize that the physical universe is not created somewhere in space at a moment somewhere in time. The big-bang universe is not created within the big bang. It is created in its entirety, equipped with a vast expanse of space and time that includes the beginning and the end. To think otherwise violates the rules of containment of the physical universe.

*   *   *

The special creation theory holds that life is miraculously created in a world already existing and previously divinely created. If the whole universe is first created as a continuum of space and time, then everything in the universe, including the origin of life, is already present in space and time. What is created once has no need to be created twice. Acts of theistic intervention in a created universe are superfluous, for everything is already created. Thus, cosmogenesis preempts miraculous creation.

Perhaps this point should be made more clear. Let us assume that God creates the physical universe. Further acts of creation are unnecessary and even irrational. A created universe is complete in every detail throughout all of space and time. Subsequent acts of miraculous creation meddle with what is already created, implying the possibility that the Creator actually exists in physical space and time, and is therefore a physical being.

Although not a science, cosmogenesis nonetheless is constrained by the logic of containment. No useful purpose is served in the physical universe by praying for theistic intervention, such as victory in battle, for the events of the past and future are created together and all inflexibly ordained.

The disparity in the estimated ages of the mythic and physical universes has vexed many persons fully convinced of the truth of the Mosaic time scale. "God could create and without doubt did create the world with all the marks that we see of old age," said Chateaubriand, a French deist. Many deists have reconciled the scriptural records with the physical evidence by supposing that God created a world already bearing the signs of great age. The zoologist Philip Gosse also saw no reason for abandoning the Mosaic chronology. In his book *Omphalos* (a word meaning navel), published in 1867, Gosse argued that Adam was created with a navel and carried the vestiges of a birth that had not occurred. If God saw fit to create Adam with a navel, said Gosse, surely God also saw fit to create a world equipped with vestiges of past eras that had in fact never occurred. The universe was created some thousands of years ago and outfitted in that cosmogenic act with a fictitious history of hundreds of millions of years. The impact of raindrops etched in sedimentary clays, footprints and teethmarks of primordial beasts, fossils embedded deep underground, light in transit from distant stars, and all the complexity of a rational self-consistent universe were created in the recent past. In the same way we might argue that the universe was created this morning before breakfast. The world comes into being equipped with a spurious past and inhabited by people with false memories. The Mosaic chronology, when applied to the physical universe, makes creation a hoax and God a jester.

*   *   *

Containment poses problems when we seek an adequate representation of ourselves in our universes. The main problem is encapsulated in what might be called the containment riddle: Where in a universe is the person thinking about that universe? This is the containment riddle that applies to the universes of all societies – Zoroastrian, Buddhist, Epicurean, Stoic, Aristotelian, Medieval, Newtonian, and so on – and not just to our modern physical universe.

The riddle is made clear by the analogy of a painter who paints a picture of his studio. A complete picture must show the studio containing the painter painting the picture. This entails an infinite regress: the picture shows the painter painting the picture that shows the painter painting the picture . . . and so on, indefinitely.

Similarly with a person in a universe thinking about that universe. The universe contains the person thinking about the universe that contains the person thinking about the universe that contains . . . and so on, indefinitely. Instead of painting a picture the person creates a mental image of the universe that, if faithful, contains in its imagery the person creating a mental image of the universe . . . and so on, indefinitely. Where in the image is the image-maker?

The answer to the riddle, where in a universe is the person thinking about that universe, is actually quite simple: the person as a conscious mind belongs to the Universe, whereas the brain – a physical representation of the person – belongs to the universe.

*   *   *

Thomas Hobbes, born at the time of the invasion of England by the Spanish Armada, wrote in *Leviathan*, "for what is the heart but a spring, and the nerves but so many strings, and the joints but so many wheels giving motion to the whole body." René Descartes followed this avenue of thought to its conclusion, and to him we attribute the total mechanization of the organic and inorganic domains. Everything in the objective world consisted of configurations of matter in motion. Give him matter and motion, and he would construct the universe. Human beings with their brains and sense organs obeyed the stern laws of the clockwork universe. The mind became a ghost haunting the machinery of body and brain. Thus began the famous Cartesian duality of mind and matter.

An inquirer might think that our Western universe has let us all down and the answer to the mind–matter problem must be found in a Taoist, Buddhist, Zen, or some quasi-physical or metaphysical universe. But the problem of the nature of mind applies to the universes of all societies, ancient and modern. Even the theologian thinking about the theocosmos must show us where in the theocosmos the theologian is thinking about the theocosmos. An image of a soul is not good enough. The theocosmos contains the soul thinking about the theocosmos that contains the soul ..., and once again we have a *reductio ad absurdum*. All who claim to know what is mind (or soul) in the context of any universe make the mistake of confusing universe with Universe.

It is of little use for the neurologist to point to his brain and say here am I thinking about the brain, because this entails a double regression. The universe contains the brain thinking about the universe that contains the brain thinking about the universe that contains ..., indefinitely. But the brain is a construct of the brain and we are caught in a regression within a regression.

Reductionists and materialists, overwhelmed by the power and majesty of the physical universe, are inclined to suppose that what is not contained does not exist. The mind is dismissed as an unnecessary fiction. Hence, the image maker is discarded leaving only the image. As in *Alice in Wonderland*, the Cheshire Cat vanishes leaving only the smile. Those who adopt this Podsnap flourish must show us where they themselves are fully portrayed in the physical universe thinking about the universe that they claim contains them thinking about the universe.

*   *   *

The containment riddle bears not only on the mind–matter issue but also on the nature of free will.

The vexed subject of individual free will versus determinism applies not only to the physical universe but to all universes. A universe makes rational the experiences of the individuals of its society. Whatever befalls a person has an explanation peculiar to that universe. The world is an activity of benign and demonic spirits; the world is dead matter jerked into motion by strings in the hands of gods; the world is a clockwork mechanism and the human soul belongs to God; the world is a dance of atoms and waves observed by bioelectrochemical brains. Always the world is lucid, rational, and deterministic.

Human beings have an awareness of free will. Yet as inhabitants of a rational universe they are components of that universe governed by its deterministic laws. They plan, but the whim of spirits controls whatever happens; they believe they are free to do this or that but in reality the *fiat* of gods determines their fortunes and misfortunes; when overwhelmed by events they are comforted by the thought that God knows best; they have the illusion of liberty to go here or there but know that all was fated long ago when the cosmic machinery started in its predestinate grooves. They labor rather than be lazy; climb mountains and go to wars rather than stay at home; join dangerous protest movements rather than rest content; study hard rather than watch television; thinking all the time that the choice is theirs, yet knowing they are at the mercy of the dancing atoms, are caught in the double helix of their genetic coding, and must follow a destiny shaped by their inheritance and environment.

We live in our individual worlds of muddled fantasy and all is redeemed and made clear by the lucidity of the universe of our society.

But the price of a universe – any universe – is that freedom of will becomes an illusion belonging to our muddled worlds of fantasy.

Marcus Aurelius, a Roman emperor and Stoic, wrote in his *Meditations*, "Whatever may happen to you was prepared for you from all eternity; and the implication of causes was from eternity spinning the thread of your being." The Stoics and early Christians devoutly believed in fate. Augustine of Hippo, architect of Christian orthodoxy, stressed in his *Confessions* the logical necessity of predestination in a universe created and ruled by God. The contrary belief that human beings have freedom of choice is the essence of the Pelagian heresy that mocks all deterministic universes. Pelagius, a British theologian and monk of the early fifth century, was appalled by the decadence of social life in Rome. His protests met only lame excuses and evasive pleas that all human weakness was preordained by God. Wickedness, he was told, being inevitable, is forgivable.

The prevailing doctrine, elaborated by Augustine and now at the heart of Christianity, declared that everything was predetermined by the will of almighty God. This was not good enough for the down-to-earth Pelagius, who lacked Augustine's conviction that the ways of God are transparent to rational inquiry. A predestinate universe threw on God the blame for wickedness that properly belonged to human beings.

Pelagius believed in freedom of individual will. He argued that the grace of God must be earned by righteous living and is a gift to all and not just to a few. He took a view contrary to that of Saint Jerome and Saint Augustine, and taught that God had not created an inalterable world of good and evil. Men and women had the freedom, if they so willed, to live untainted by sin, inasmuch as God had given them that freedom. "If it is necessary, then it is not a sin," he said, "if it is optional, it can be avoided."

There was much dispute and Augustine won the battle with the aid of scriptural testimony. A rational universe, controlled by divine will, in which human beings behaved as robotic creatures, triumphed.

The Pelagian defense of free will was condemned as heresy in AD 418.

*　　*　　*

The Universe contains us devising universes. The conscious mind and its free will belong to the Universe; the body, brain, and determinism belong to the universes.

16 *Ultimum Sentiens*

I see the world around,
So simple yet profound.
Specifically, I see a tree.
The retinas of my eyes,
– sensitive optically –
Do not consciously
See actually the tree.
Signals to my brain
Electrobiochemically
Produce synaptically
A connected pattern
That represents the tree.
But what sees the pattern?
Pray where neurologically
Is the *ultimum sentiens*
That with full attention
Really is *me* consciously
Seeing the tree?
Is my understanding
Of me seeing the tree
A neurological pattern
That explains me
Seeing the tree
As a neurological pattern?

Consciousness is a nonphysical property that cannot be defined
in physical terms, and indeed does not exist in the physical universe.
It is impossible to determine by any physical means if an object is

conscious. When presented with miscellaneous objects, such as an orange, a chair, a clock, a human being, a candle flame, and a crystal, an experimenter cannot determine by means of experiments with physical equipment which of these objects is conscious. This normally would constitute sufficient proof that consciousness does not exist anywhere in any form. One of the objects, however, could be myself, and I know beyond all doubt that I am a conscious being. I am more certain of my consciousness than was Dr. Johnson of the concreteness of his stone. Consciousness beyond all doubt exists, yet demonstrably does not exist in the physical universe. Consciousness belongs to the Universe not the physical universe. No other conclusion seems possible.

That consciousness exists is experimentally an unfalsifiable fact. Yet all knowledge arises from conscious experience. We are aware of its existence because it is self-aware. With a Podsnap flourish many thinkers nonetheless consciously deny that consciousness exists. Some inquirers have supposed that consciousness is a sort of vitalistic property that emerges in complex neurological systems. Such a property, however, must therefore be physical and amenable to experimental verification.

* * *

In the spirit of the exact sciences we study the brain and observe it as an objective entity; we study its neurological structures and their functions, and endeavor to discover how the brain works. We avoid confusing the inquiry by disavowing psychological and theological terms such as consciousness and soul that are objectively unmeasurable and physically meaningless. We stick to one outlook (the only rational outlook in the physical world) and the logic of one language (that of the exact sciences). We exist in a colorless, toneless, tasteless, scentless, unfeeling shadowy theoretical world of atoms and electromagnetic and gravitational fields.

Much of the brain is still *terra incognita* and we have yet to understand fully the physical basis of memory, sleep, and even things like headaches. But nobody doubts that understanding will come in the future.

Signals carrying information from outside the physical body impinge on the sense organs and are relayed through neuron pathways to the brain. The torrent of incident information in the form of light rays, sound waves, and other signals greatly exceeds what can be handled, and hence the sense organs, intermediate structures, and pathways filter and regulate the incoming flow of information. The retina and visual cortex, for example, have their interconnected clusters of cells responding selectively to movements and shapes, to edges and lines of specific orientation, thereby enhancing significant features in the visual field. The brain consists of tens of billions of neurons; each has radiating fibers linking it to thousands of nearby and distant neurons, and the brain with its ten trillion linkages acts as an omniconnected computer network that receives, formats, relates, creates, transmits, and stores information.

We can imagine the brain as a hierarchy of neurological structures. At the highest levels in the cerebral cortex are constellations of interconnected neurons that account for active thought, speech, short-term memory, and many of the characteristics of individual behavior. Here and at lower levels exist the soft-wired structures programmed by language, culture, and personal experience. The hard-wired structures that predominate at lower levels and are linked with the glandular chemistry of emotional response operate in genetically programmed ways.

Colin Blakemore in his BBC Reith Lectures *Mechanics of the Mind* said, "The study of the brain is one of the last frontiers of human knowledge and of more immediate importance than understanding the infinity of space and the mystery of the atom." The study of the objective brain is undoubtedly of vital importance, for without knowledge of how the brain works, how can we, who are brains according to this outlook and language, have confidence in the brain's reconstruction of the objective world, which includes the brain itself? Blakemore concluded his lectures with the words, "The brain struggling to understand the brain is society trying to understand it-self." At present in our modern universe we are confronted with the problem of understanding rocks and trees, and the brain reconstruct-ing a universe of rocks and trees and the brain.

We live in an age of science and are all scientists or, at least, de-pendent on the products of science. Philosophy and theology, which once sought to explain the human condition, are left far behind en-meshed in the coils of ancient belief systems on which science casts a frosty eye. The body–mind problem has been dismissed, and the mind, it is said, is no more than another name for the brain.

But there is a problem. The brain studies the brain interact-ing with the physical world of which it is a part; this picture of the brain studying the brain is a construct of the brain itself. There is a regression of brains studying brains constructing pictures of brains studying brains constructing... Plausibly, the solution requires that we recognize that more than one kind of brain is involved. The ob-serving brain studies the objective brain in a picture constructed by a conscious brain. We might try to say the brain constructing the picture is a robot. But a robot constructing a picture of the brain studying the brain is an imaginative picture constructed by the conscious human brain. The mind creeps back as the *ultimum sentiens* in the form of consciousness. The conscious mind that belongs to the Universe studies the brain that belongs to the universe.

*   *   *

What distinguishes the observer from the observed? Where lies the demarcation between the observer and the observed? In an observa-tion the observer interacts with the observed object and forms a joint physical system. The observer–object interaction is no more than an object–object interaction in the physical world. Consciousness, a non-physical property of the observer, plays no role in the physical interac-tion. For the observation to become a conscious experience, a second observer must observe the first who has become absorbed into the object under observation. But the second observer forms also a joint physical system in which consciousness again plays no physical role. Hence there must be a third who observes the second, a fourth who observes the third, and so on, indefinitely.

Where lies the *ultimum sentiens*? A proposed way of terminating endless regression hypothesizes an ultimate observer of sufficient complexity to generate the psychic properties of an *ultimum sentiens*. This approach fails because the required property of the *ultimum sentiens* is consciousness, which we previously have shown has no existence in the physical universe.

In desperation we might suppose a conscious homunculus or mannikin occupies the brain and is responsible for our acts of perceiving. But again we run into an infinite regression of homunculi occupying homunculi.

The observer–observed problem restates the Cartesian mind–body problem. Both address the issue of the mind consciously thinking about and observing objects in the physical universe. The problem has no solution other than to deny that the mind is more than the brain, a so-called solution that lapses into infinite regress.

The conscious *ultimum sentiens* is important in understanding the nature of observations in quantum theory. In the quantum world, atomic systems are represented by virtual wavefunctions that collapse into real states of known probability whenever an observation occurs. The quantum virtual world consists of evolving wavefunctions that constantly branch into new wavefunctions, thus forming a manifold of many potential worlds, all coexisting, all theoretically deterministic, all having probabilities of being the actual world into which the phantom wavefunction manifold collapses when an observation occurs. The wavefunction manifold evolves in a deterministic manner, but its collapse into definite states is always an indeterminate event. The observer forms part of the wavefunction manifold in an observer–object interaction, or more usually, an observer–apparatus–atomic system interaction. The atomic system interacts with the observer–apparatus system and their combined wavefunctions represent two objects interacting with each other. Nothing in this description can cause the wavefunction to collapse into an observation. Seeking the answer we invoke a second observer. But this enlarges the interacting systems, thus adding further components to the wavefunction, and yet still no observation occurs because nothing in the representation causes the collapse of the wavefunctions. So we invoke a third observer, and then a fourth, and so on, in search of the *ultimum sentiens*. The impotency of the wavefunction to collapse into a definite observed state is known as von Neumann's regression. At each step we seek termination by unsuccessfully invoking a new observer. The phantom world, lacking an *ultimum sentiens*, never collapses into the definite states of the observed world.

Problems of this nature abound when we divide the physical world into the observer and the observed. Their solution requires, so it seems, that the observer as an *ultimum sentiens* possesses consciousness and is more than a physically interacting system.

We cannot point to the brain with its neural networks and say "here is the conscious mind." For in its cerebral activity lies a representation of the cerebral activity that is us saying "the conscious mind is the brain." The brain studying the objective world containing the brain is a representation that is itself a neural pattern. We have thus a picture of the interplay of neurons representing the interplay of neurons representing . . . and so on, indefinitely. This is, in effect, a picture of the world at which nobody is looking. This is not unlike von Neumann's endless regression of wavefunctions that never collapse into an observation for want of a conscious observer.

*   *   *

The *Tale of a Danish Student* by Poul Møller is an account of a student searching for his real self. The student thinks,

> . . . man divides himself into two persons, one of whom tries to fool the other, while a third one, who in fact is the same as the other two, is filled with wonder at this confusion. In short, thinking becomes dramatic and quietly acts the most complicated

plots with itself and for itself; and the spectator again and again becomes actor.

The more deeply the student delves within himself the more elusive becomes his real self:

> ... then I come to think of my thinking about it; again I think that I think of my thinking about it, and divide myself into an infinitely retreating succession of egos observing each other. I don't know which ego is the real one to stop at, for as soon as I stop at any one of them, it is another ego again that stops at it. My head gets all in a whirl with dizziness, as if I were peering down a bottomless chasm, and the end of my thinking is a horrible headache.

Niels Bohr was much impressed in the early days of quantum mechanics by this tale of self-discovery. The problem of the *ultimum sentiens* lies at the roots of quantum theory, and indeed of much of the philosophy of science.

On a similar subject Erwin Schrödinger, a pioneer in the formulation of quantum theory, wrote in *Mind and Matter*:

> Sometimes a painter introduces into his large picture, or a poet into his long poem, an unpretending subordinate character who is himself. Thus the poet of the *Oddysey* has, I suppose, meant himself by the blind bard who in the hall of the Phaeacians sings about the battles of Troy and moves the battered hero to tears. In the same way we meet in the song of the Nibelungs, when they traverse the Austrian lands, with a poet who is suspected to be the author of the whole epic. In Dürer's *All Saints* picture two circles of believers are gathered in prayer around the Trinity high up in the skies, a circle of the blessed above, and a circle of humans on the earth. Among the latter are kings and emperors and popes, but also, if I am not mistaken, the portrait of the artist himself, as a humble side-figure that might as well be missing. To me this seems to be the best simile of the bewildering double role of mind.

Elsewhere in *Mind and Matter* Schrödinger wrote, "I so to speak put my own sentient self (which has constructed this world as a mental product) back into it – with the pandemonium of disastrous logical consequences." And Alexandre Koyré in *Newtonian Studies* remarks, "This is the tragedy of the modern mind which 'solved the riddle of the universe,' but only to replace it by another: the riddle of itself." The "riddle of itself," for me at least, is substantially solved by realizing that the observer is a conscious being, that the prime characteristic of the *ultimum sentiens* is consciousness, and that consciousness does not exist in the physical world but in the Universe that lies behind the eyeless masks that are our universes.

*　　*　　*

The brain throbbing with ceaseless electrochemical activity is part of the physical world, and though much is still not understood, rapid progress is being made. The brain and computer have much in common, and developments in computer science help us to understand more about the brain. Through their input channels the brain and computer receive information that is modified, processed, stored, applied, and transmitted. Both respond in elaborate ways to their environments. It does not take much imagination on our part, a little courage only, to foresee a time when artificial intelligence will rival and then outrival human intelligence in all applications. All this is within the context of the physical world description.

Human beings are by no means perfect. With their muddled and emotional thinking they seem incapable of making clear-sighted, far-seeing decisions on which our survival depends. The idea of surrendering the day-by-day control of our political and judicial systems to pondering logical machines of advanced artificial intelligence does not bother me; sometimes I think they are the one ray of hope for a sane future. Modern society is far too complex for the human brain to compass and regulate in an orderly and sensible fashion. The master machine-minds of the Earth conferring together would know in microseconds that nuclear and biochemical weapons are not in the interests of their human populations, and in milliseconds would agree on how to eliminate them. They could optimize the size of human populations, humanize all aspects of society, organize healthcare and education, husband natural resources, and administer effective laws that minimize the misery of incarceration. The aim? To enhance the joy and quality of life without undermining our morality and destroying the environment. If Utopia lies ahead in the history of the human race, it will be attained by the computer in control of the machinery of society.

The notion of artificial intelligence outrivaling human intelligence in all cerebral functions is abhorrent to most persons. These slavelike computing creatures of the physical world must at all costs be kept in their places. We hide behind an assumed monopoly of intelligence and protest when anyone attempts to measure it. I am tempted to think that what the brain can do the machine will eventually do very much better.

Many of our fears of artificial intelligence stem from the belief that the physical universe is the Universe. Hence, the physical universe contains all that exists, and all that exists is necessarily physical. At all costs we must preserve our status as superior physical entities among all physical things. But we have no need to fear; we are the conscious devisors of the physical universe in which non-conscious artificial intelligence exists.

*　　*　　*

It would be difficult to imagine any experience consisting of sights, sounds, thoughts, and emotions that is not conscious. We are aware of our experiences and emotional and mental states because of consciousness. We see the distant blue hills and the birds flying overhead, we feel the caress of breezes, smell the scent of verbena from the garden, hear the chuckle of a distant coyote, and we sense the mystery of it all; we suffer grief; we are uplifted by song, the movement of dance, a memorable painting, and all are fragments of the stream of conscious experience.

Why are cognitive processes in the brain accompanied by consciousness? Could not a robot, an unconscious automaton, perform the same tasks just as well? Perhaps. Perhaps even better. But it would be a dead thing with none of the joys and griefs of conscious life. In a profound way we have come full circle and discovered what the primitive person always knew: we are touched by divinity, the divinity of an undefinable thing called consciousness.

17 All That is Made

In the tenth and eleventh centuries the Arab dialecticians of the *Kalam* (the *Mutakallimun*) were opposed to the Aristotelian science in Muslim theology and professed a theory of extreme theism. Everything, they said, is governed by the will of the Sole Agent. They exalted the power of the Sole Agent by squeezing dry the natural world of all ability to be self-explanatory. The *Mutakallimun* devised their own interpretation of the atomism of the Epicureans. The *Kalam* atoms were completely isolated and noninteracting. Not only matter but space also was atomized. Nothing bridged the atomic gulfs except the harmonizing and coordinating power of the Sole Agent.

Al-Bakillani, a disciple of a disciple of the famed al-Ashari (the founder of Muslim scholasticism), lived in Baghdad where he died in 1013. He introduced the idea of atomic time. His ideas were critically discussed in the twelfth century by Moses Maimonides in *The Guide for the Perplexed*. "An hour," explained Maimonides, "is divided into sixty minutes, the minute into sixty seconds, the second into sixty parts, and so on; at last after ten or more successive divisions by sixty, time-elements are obtained, which are not subjected to division, and in fact are indivisible, just as is the case with space."

In al-Bakillani's scheme, atoms divided up space and time. The universe, without continuity in space and time, manifestly was under the coordinating control of the Sole Agent. In an atom of time the universe exists in a fixed state of being. The state of being dissolves and in the next atom of time a new state of being is created in a slightly different state. The universe is created not once, but repeatedly. God was the producer and director of a cinematographic universe.

*   *   *

Julian of Norwich in the fourteenth century wrote in her *Revelations of Divine Love*, "I saw that he is everything that we know to be good."

> And he showed me more, a little thing, the size of a hazelnut, on the palm of my hand, round like a ball. I looked at it thoughtfully and wondered, "What is this?" And the answer came, "It is all that is made." I marveled that it continued to exist and did not suddenly disintegrate; it was so small. And again my mind supplied the answer, "It exists, both now and forever, because God loves it."

Redeemed by her Mother Jesus, the universe was the temple of her Mother God. Her reverence, not unlike that of the *Mutakallimum*, was for the Creator and not the created.

Four centuries later in the Age of Reason, the theism of al-Bakillani and Julian of Norwich, the theism of all that is made and sustained, transformed into deism, the deism of all that is made and self-sustained. Reverence of the Creator was in part transformed into reverence of the created. In "The tables turned" in the *Lyrical Ballads* (1798) William Wordsworth wrote:

> One impulse from a vernal wood
> May teach you more of man,
> Of moral evil and of good.
> Than all the sages can.

The natural world had begun to recover what for so long had been lost.

*   *   *

William Paley, archdeacon of the diocese of Carlisle and a strenuous supporter of the abolition of slavery, spent much time in his study where he tried to compensate for his "deficiencies as a churchman." He is celebrated for his *Natural Theology*, published in 1802, three years before his death. The subtitle *Evidences of the Existence and Attributes of the Deity Collected from the Appearances of Nature* summarized the theme of his culminating work.

Notice, wrote Paley, how well-contrived is the eye wherein all parts cooperate to serve a common purpose, and similarly the hand. Never in all eternity could the eye and hand have arisen by themselves in response to the blind forces of nature. Clearly, all things of the living world were designed by an intelligent deity expressly for the purposes they ably fulfill.

Suppose, wrote Paley, that while walking on the heath I stumble against a stone. I would not feel provoked into wondering how the stone got there, for it may have lain on the ground for untold ages. "But suppose I found a watch on the ground, a natural conclusion would be the watch must have a maker; that there must have existed at some time and at some place an artificer or artificers who formed it for the common purpose which we find it actually to answer, who completely comprehended its construction and designed its use." All around us we see intelligent design "such as relations to an end, relations to one another, and to a common purpose," and wherever we witness the formation of things, the evidence of God the designer, the clockmaker, stares us fully in the face.

Paley ably expressed the views of the deists. More than a century previously the clergyman–scientist Thomas Burnet had anticipated much the same in his *Theory of the Earth*: "We think him a better Artist that makes a clock that strikes regularly at every hour from the springs and wheels he puts in the work, than he that hath so made his clock that he must put his finger in it every hour to make it strike." Deists in the eighteenth and nineteenth centuries often used the clock analogy. The closer they studied the clockwork mechanism, the more obvious seemed the evidence of intelligent design. As theistic maintenance waned with the advance of science, deistic design waxed.

Unfortunately, William Paley's wonderment at all that is made contributed nothing to explaining how the mechanisms worked.

Deism was left with no recourse but to peer deeper into the machinery to find out how it operated. More and more the design was found to be the consequence of natural processes. The stone was no less wondrous than the watch. Water seemed as much purposive in its properties as the eye and hand. If water did not expand on freezing – and ice did not float – the oceans would freeze solid and life on Earth would become impossible.

In the *Bridgewater Treatise*, written by eight distinguished authors and dedicated to demonstrating the "Power, Wisdom and Goodness of God as manifested in the Creation," the chemist William Prout wrote in 1834, "The above anomalous properties of the expansion of water and its consequences have always struck us as presenting the most remarkable instance of design in the whole order of nature – an instance of something done expressly and almost (could we indeed conceive such a thing of the Deity) at second thought to accomplish a particular object." Little by little the structures of the living and non-living worlds were found to be implicit in the basic properties of matter. The atomic blueprint was designed in a way that atoms worked naturally to form the wonders of the world.

Paley saw evidence around him of an Artificer who fashioned inert matter into elaborate structures. But the universe of Paley's day was changing. Design became apparent less in the particulars, more in the general, less in the eye and hand, more in the molecular components. Design operated through natural selection, the laws of the heavens, and the fabric of space and time. This is where we stand today. Most intellectuals believe the universe in which we live is self-running, some even think it is self-creating.

*　　*　　*

Why does the universe exist? A popular answer is because God made it. But why only this one and not many others? Giordano Bruno's argument in 1584 of theistic plenitude can be extended to the creation of a multitude of universes: "Thus is the excellence of God magnified and the greatness of his kingdom made manifest; he is glorified not in one, but in countless suns; not in a single earth, but in a thousand, I say, in an infinity of worlds." Perhaps other universes were preliminary experiments performed before the construction of our own. David Hume in *Dialogue Concerning Natural Religion*, published in 1779, wrote that numerous universes "might have been botched and bungled throughout an eternity ere this system was struck out; much labour lost, many fruitless trials made, and a slow but continual improvement carried out during infinite ages in the art of worldmaking."

Why stop at our universe? Might not other universes thereafter have been struck out more splendid than the one we now inhabit? Olaf Stapledon in his imaginative book *The Star Maker*, described in 1937 how the Star Maker created universes of increasing magnitude and complexity until each far surpassed our own:

> In vain my fatigued, my tortured attention strained to follow the increasingly subtle creations which, according to my dream, the Star Maker conceived. Cosmos after cosmos issued from his fervent imagination, each one with a distinctive spirit infinitely diversified, each in its fullest attainment more awakened than the last; but each one less comprehensible to me. ... I strained my fainting intelligence to capture something of the form of the ultimate cosmos. With mingled admiration and protest I haltingly glimpsed the final subtleties of world and flesh and spirit, and of the community of those most diverse and individual beings, awakened to full self-knowledge and mutual insight. But as I strove to hear more inwardly the music of concrete spirits in countless worlds, I caught echoes not merely of joys unspeakable, but of griefs inconsolable.

Some of Stapledon's universes consisted solely of psychic phenomena, yet others solely of physical phenomena, but most combined both. Some universes formed clusters whose members were either interconnected or totally isolated from one another. In "one inconceivably complex cosmos,"

> ... whenever a creature was faced with several possible courses of action, it took them all, thereby creating many distinct temporal dimensions and distinct histories of the cosmos. Since in every evolutionary sequence of the cosmos there were many creatures and each was constantly faced with many possible courses, and the combinations of all their courses were innumerable, an infinity of distinct universes exfoliated from every moment of every temporal sequence in this cosmos.

The "exfoliating" universe introduced an interesting new twist in cosmological thinking.

$$*\quad*\quad*$$

"Time forks perpetually toward innumerable futures," wrote Jorge Luis Borges, Argentinean essayist and connoisseur of the bizarre, who pursued the idea of the exfoliating universe in *The Garden of Forking Paths*. The concept of forking time provides an imaginative solution to the age-old problem of free will versus determinism, but at the cost of invoking multiple branching universes.

Pure chance and freedom of choice are unwelcome guests in any rational (i.e., deterministic) scheme of things. As in an authoritarian society, what is not mandatory is forbidden. Imagine that Mr. Smith when walking in a wood comes to a place where the path divides into two paths. Suppose that there is no reason why he should take one path more than the other and he is free to choose which he pleases. But freedom of choice is an illusion in a rational universe for all is determined by the laws governing that universe. Liberty to do this or that as you wish and go here or there as you please is an illusion. Hence Mr. Smith, having no free will, like Buridan's ass, does nothing. But this is like Zeno's paradox; we know that Achilles overtakes the hare, and we know that Mr. Smith makes a choice. Not able to choose one path more than the other, he takes both paths.

How is this possible? He cannot be at two places at the same time, therefore, as in Stapledon's cosmic exfoliation, he takes each path in a different universe. The universe splits at each indeterminate situation.

"Time forks perpetually toward innumerable futures," said the essayist. Every indeterminate situation is resolved by realizing all possibilities in different universes. What is potential always becomes actual. When the laws of nature are impotent the universe divides. This is the remarkable many-worlds denial of indeterminism. The exfoliating universe applies to all situations in which there is no reason for one thing to happen more than another. At each indeterminate event the universe splits into several universes in which all possible outcomes are separately realized.

*　　*　　*

In 1957, some years after Stapledon wrote the *Star Maker*, the many-worlds argument was introduced into physics by Hugh Everett at Princeton University. We can imagine the atom as a bundle of waves. When disturbed it consists of evolving waves representing the various possible final states. If an observation is made the wave picture collapses into a particle picture that realizes only one of the possible final states. The observer never knows in advance what the final state will be and can only predict its probability from the waves. The theoretical world of many waves – potential of many futures – is fully deterministic; the observed world of particles is uncertain and its future can be predicted only with probability.

"God does not play with dice," said Albert Einstein, who was opposed to this picture of future uncertainty, and thought there should be a more fundamental deterministic theory. One way of recovering determinism at the atomic level is the many-worlds interpretation of quantum mechanics. Instead of an uncertain final state of known probability, the atom realizes all possible states, each state in a different universe. Thus, the atom of many potential futures actually realizes all states in different universes.

At each quantum transition the universe splits. Bryce DeWitt, who has contributed to the many-worlds theory, remarks, "every quantum transition taking place on every star, in every galaxy, in every remote corner of the universe is splitting our local world on Earth into myriads of copies of itself. I still recall vividly the shock I experienced on first encountering this multiworld concept." The number of exfoliating universes is enormous. As an illustration: the number of atoms in the visible universe is roughly the Eddington number 10^{80} (10 followed by eighty zeros); if we suppose every atom makes one transition each second, our universe generates 10^{80} universes each second, and if we assume a lifetime of the universe of 10^{20} seconds, we find our universe generates a googol of universes. (1 googol $= 10^{100} = 10$ followed by 100 zeros.) Each member of this googol ensemble generates itself a googol ensemble of universes of which each generates a googol ensemble of which each generates ..., and so on.

* * *

Why is the universe designed to be compatible with the existence of life? Why is it organized with planets, stars, and galaxies, furnished with laws of a certain nature, and equipped with fundamental constants (such as the mass and charge of the electron) of particular values? These questions are not scientific in the ordinary sense but are more cosmological and even philosophical. They tend also to be theological because the answer often given is that God designed the universe specifically for inhabitation by life.

The question "why does the universe exist?" is not the same as the question "why is the universe the way it is?" Nor are the answers (if answers there be) necessarily the same. The first question relates to the existence of the universe and the second to its compatibility with the existence of life. From the theological point of view the subjects of cosmic creation (cosmogenesis) and cosmic design (fitness) are intimately related. From a scientific point of view, however, the subjects of creation and fitness involve very different issues and recognition of this difference facilitates rational inquiry.

That God created and also designed the universe is a possible argument. Another, an Aristotelian kind of argument, revived in recent years, is now known as the anthropic principle. Lawrence Henderson, a scientist of broad interests at Harvard University, wrote in *The Fitness of the Environment* in 1913:

> The fitness of the environment results from characteristics which constitute a series of maxima – unique or nearly unique properties of water, carbonic acid, the compounds of carbon, hydrogen, and oxygen, and the ocean – so numerous, so varied, so nearly complete among all things which are concerned in the problem that together they form certainly the greatest possible fitness.

A "fit" universe is a universe fit for inhabitation by life. Because life exists, the universe is necessarily designed the way it is. It is the old design argument with a new twist in which the deity need never be mentioned.

Life exists not because the universe by chance or intention is a fit place for habitation, but the universe necessarily is a fit place because it contains life. If by mischance the universe were unfit, we would not be here to comment on its unfitness. Our existence places tight constraints on the nature of the universe. The deistic argument has been turned upside down. We must postulate the existence of human beings, not God, if we wish to understand why the universe is necessarily the way it is. This is the essence of the anthropic principle that inverts the deistic design argument.

* * *

The basic properties of the universe appear to be arbitrarily determined and fixed in ways we do not understand. In Aristotle's terminology they are the "accidentals" of this world. The speed of light, the strength of gravity, Planck's constant of quantum theory, the electric charges and masses of atomic particles, and other constants such as the strengths of the basic interactions are all not determined uniquely by any known theory and appear to be accidental. They even appear to be providential because if they were different in value, even very slightly, we would not be here discussing the subject.

Let us suppose there are many physical universes, each complete and self-contained in its own space and time. We may suppose, if we wish, that they all occupy a superspace of some kind, but nonetheless isolated and noninteracting. Among these many universes the accidentals – the inexplicable constants of nature – are distributed with various values and arranged in all combinations. Gravity is stronger in some and weaker in others than in our own; in some the electric charge of the electron (and proton) is larger and in others smaller; and similarly with the rest of the constants. Each cosmos in the multiuniverse serves as a workshop in which we examine the consequences of the accidentals having values other than in our own. Study of this ensemble of universes leads to a truly astonishing conclusion.

We find that most universes contain only hydrogen. The nuclei of atoms heavier than hydrogen cannot exist (because the electric repulsion between protons is too great or the strong interaction between nucleons is too weak) and these universes lack elements necessary for the formation of planetary systems and biological organisms. They lack in particular the carbon, nitrogen, and oxygen necessary for organic molecules and the biochemistry of life. Living creatures, as far as we know, cannot be constructed from hydrogen only, and these hydrogen-only universes are lifeless.

On looking closer at the ensemble we find among the universes capable of having stable heavy elements that many are without planets, stars, and galaxies and consist only of a featureless distribution of gas. In these "grin universes," where "all nature wears one universal grin" (Henry Fielding), we find for various reasons (gravitation is too weak, temperature too high, expansion too rapid, cosmic lifespan too short) that the conditions are unfavorable for the formation of astronomical systems.

In many universes having astronomical systems the stars are cold and dark. Throughout these inhospitable world systems of perpetual darkness the stellar furnaces remain unlit and the industry of producing heavy elements from hydrogen and helium stands idle. In only very few universes the stars shine brightly. Fewer still have luminous lifetimes of billions of years, long enough for biological evolution to occur. Stars must not evolve too slowly, nor too rapidly and burn out before life originates and evolves to complex states. Examination of the whole ensemble reveals that life exists probably in only one universe – the one we inhabit – or in a small fraction of universes almost indistinguishable from our own.

*   *   *

Why is our universe the way it is? One answer is because we exist. This is the anthropic principle, named by Brandon Carter, who has explored the fitness of the cosmic environment as a "reaction against exaggerated subservience to the Copernican principle." The Copernican (or rather Democritean) principle asserts that human beings rank as inconsequential incidents in the scheme of things.

Usually, scientists translate existential questions into functional questions. Ontological perplexity is not in their department. "Why is the universe the way it is?" translates into "how has the universe evolved?" which means finding the appropriate initial conditions and the governing laws. But now they have a new answer, the anthropic principle, in which an ontological question receives an ontological answer. The principle states that the universe is the way it is because we exist. The usual question, "why does life exist?" with the usual answer, "because the universe is the way it is," has been turned around and becomes, "why is the universe the way it is?" with the anthropic answer, "because life exists." All other universes of different construction lack the essentials of life, such as long-lived luminous stars and elements heavier than hydrogen, and hence are lifeless. Only a very small subset of universes contain living creatures because the accidentals have come together with precisely the right values requisite for life.

One might speculate on the possibility of a consciousness principle: only the universes are real that contain at some time cognizant conscious inhabitants. Without conscious inhabitants a universe is only virtual and never real. The anthropic principle raises the specter of a wasteland of lifeless universes; the consciousness principle dismisses the wasteland as virtual, leaving only the conscious universes as real.

*   *   *

Design is either fortuitous or intentional. From an Aristotelian viewpoint we see the cosmic design parameters as fortuitous and accidental in origin. Their randomness implies that the stupendous ensemble of universes may actually exist. The parameters have in a haphazard manner come together in our universe with the precise values requisite for the origin and evolution of life. Countless universes, impotent to spawn life, are plunged in total darkness or filled with searing light.

Alternatively, design is intentional. From a theological viewpoint we see that our finely tuned universe was designed by God specifically for inhabitation by life. The universe contains swarms of galaxies, oceans of space and time, long-lived luminous stars, and finely adjusted constants of nature in order that life shall exist. Granted that the universe is the way it is because life exists, but it is made precisely this way and no other in order that life shall exist. The deistic principle (God created and designed the universe) has no need of a wasteland of barren universes, and the whole ensemble may be discarded as a theoretical fiction serving the purpose of demonstrating the fitness of the universe we occupy. Why is the universe the way it is? Because God made it that way and no other in order that life shall exist. Here is the cosmological proof of the existence of God – the design argument of Paley – updated and refurbished.

*   *   *

Intentional design implies intelligence. In that case why not invoke the intelligence of "angels" rather than of God – the intelligence of conceivable beings rather than of an inconceivable supreme being? Perhaps our universe was created by life of superior intelligence existing in another universe in which the finely tuned constants of nature were compatible with the existence of life, and therefore essentially similar to our own.

According to Alan Guth we already know how in principle universes can be made. The trick is to form a small black hole with its interior conditions at the precise point for the onset of inflation. The interior space inflates, creating matter and forming a vast separate universe. This possibility suggests that intelligent beings, including our own descendants in the far future, might possess not only the knowledge but also the technology to design and manufacture universes. We have thus the basis of a theory of natural selection of universes. Intelligent life in a parent universe creates universes, and in the offspring universes fit for inhabitation new life evolves to a high level of intelligence and creates further universes. Universes unfit for inhabitation lack intelligent life and cannot reproduce. Plausibly, offspring universes have properties that are closely similar to their parent universes – apart from small genetic variations in the constants of nature – and the universes most hospitable to intelligent life are naturally selected by their ability to reproduce.

Why is our universe the way it is? Because, according to this theory, it is similar to the parent universe from which it springs, which contains life, and is already finely tuned. Why is the universe comprehensible to the human mind? Perhaps because it was created by comprehensible beings of finite intelligence. Why make universes? Perhaps the final goal of all intelligent life is to create universes, habitable yet diverse, that can be explored and colonized by their creators.

*   *   *

When given a hazelnut we share Julian of Norwich's wonder of all that is made, a wonder now more enlightened by our knowledge of a universe of vast expanses of space and time, of multitudinous galaxies, stars, and planets.

It is all very well for us to adopt a god's-eye view while surveying the creation and design of universes, as in Stapledon's extracosmic theater, but with our ordinary worm's-eye view how can we ever verify that any of these other universes actually exists? Each is self-contained and beyond observation by human beings. The assertion in Francis Thompson's *Kingdom of God*: "O world invisible, we view thee, O world intangible, we touch thee, O world unknowable, we know thee," is fine for the poet, but of no help to the cosmologist. When postulating other universes we quit the solid ground of empirical knowledge for the airy heights of unfalsifiable speculation. As Hume said on this subject, "who can determine where the truth . . . lies amidst a great number of hypotheses that may be proposed and a still greater that may be imagined?"

Have we once more failed to distinguish between our universe and the Universe? All universes, Aristotelian, Medieval, Newtonian, . . . , the modern physical universe, and the universes of the future are representations of an underlying reality as understood by the human mind. In no circumstances may we imagine the Universe itself as a member of an ensemble of Universes, for the Universe is inconceivable and patient of many representations.

The anthropic and natural selection principles address the subject of cosmic design but leave aside the question of cosmic creation. They do not explain the origin of an ensemble of universes. The deistic principle addresses the subject of creation but begs the question of who created God. If we conjecture that God is self-creating, why should we not ascribe this property directly to the Universe? The many universes, singly and in ensembles, are creations by the human mind in its quest to understand the Universe.

18 The Cloud of Unknowing

An unidentified English author of the fourteenth century, who was probably a priest, wrote

> But now thou askest me and sayest: "How shall I think . . . and what is he?" Unto this I cannot answer thee, except to say: "I know not." For thou hast brought me with thy question into that same darkness, and into that same Cloud of Unknowing. . . . For of all other creatures and their works – yea, and of the works of God himself – may a man through grace have fullness of knowing, and well can he think of them; but of God himself can no man think. And therefore I would leave all that thing that I can think, and choose to my love that thing that I cannot think.

Like other contemplative mystics of the Middle Ages the author discovered that thought could not unveil the face of God: "By love may he be gotten and holden; but by thought neither." God, the Cloud of Unknowing, was beyond articulation, and the source of all articulations.

Contemplative mystics in the Middle Ages – Christian, Jewish, and Muslim – ranked among the most advanced thinkers of their time. Thus, Nicholas of Cusa, prince and statesman of the Roman Church, sagely recorded that "scientific superstition" is the expectation that science answers our every question.

* * *

In the West, and wherever else the modern physical universe now holds sway, sections of the public have caught up with the agnostic intellectuals of the nineteenth century. The educated person now finds it difficult not to be agnostic. Agnosticism is the belief that the existence of God may be affirmed by faith but not by appeal to reason.

The word agnostic, often misunderstood, was first used by Thomas Huxley at a party in London one evening just before the founding of the Metaphysical Society in 1869. A few months later the *Spectator* reported that Huxley "is a great and even severe Agnostic who goes about exhorting all men to know how little they know." In a subsequent issue of this journal we learn, "Agnostic was the name demanded by Professor Huxley for those who disclaimed atheism and believed with him in an 'unknown and unknowable' God; in other words, that the ultimate origin of things must be some cause unknown and unknowable." The author of *The Cloud of Unknowing* was an agnostic. Whereas atheists deny the existence of God, agnostics accept the possibility, and following Huxley, deny that God is known and knowable.

To avert the snare of agnosticism and give comfort to all of faltering faith, the Vatican Council decreed that "man can know the one true God and Creator with certainty by the natural light of human reason." Natural theology is the branch of cosmology that aims with "the natural light of human reason" to find evidence of God's existence. Agnosticism disputes the assertion that God is knowable and known by reason alone. Both natural theologians and agnostics claim to use reason alone and exclude faith from their discussions.

Natural theology began with the Ionians and entered the mainstream of Christian thinking in the Middle Ages. Saint Anselm in the eleventh century, who was archbishop of Canterbury for the last sixteen years of his life, is the first known Christian to attempt to prove the existence of God by means of pure reason independent of religious belief.

The proofs of God's existence fall into four main groups, referred to as the ontological, moral, cosmological, and teleological arguments.

The ontological argument seeks to show that the existence of God can be demonstrated by propositions of indisputable truth and that the reality of God is the prime essential of all reality. The moral argument seeks to demonstrate that without God there would be no certain distinction between good and evil. The cosmological argument seeks to show that the universe could not exist without its creation and maintenance by God. The teleological argument seeks to show that the universe is designed to serve definite purposes and attain specific ends.

A miscellany of proofs, devised by Thomas Aquinas and referred to as the Five Ways, illustrates the thrust of natural theology:

> Things are in motion, hence there is a first mover.
> Things are caused, hence there is a first cause.
> Things exist, hence there is a creator.
> Perfect goodness exists, hence it has a source.
> Things are designed, hence they serve a purpose.

At the end of each proof Aquinas added, "all understand that this is God," or words to that effect. The first three relate to the cosmological argument, the fourth relates to the moral argument, and the fifth to the teleological argument. Aquinas had no patience with the ontological argument and thought it logically unsound.

* * *

The ontological argument was initiated by Anselm, who hit on the idea of defining God as "that being than which nothing greater can be conceived." He argued that the reality of God is greater than the idea of God, and therefore by definition, God exists.

Anselm was delighted with his proof. But it misfires, for what is conceived by the mind is not necessarily the truth. Almost two centuries later, Aquinas rejected Anselm's ontological argument and said, granted that the supreme being can be so defined, "it does not follow that what the name signifies actually exists, but only that it exists mentally." Parodying Anselm, we could define Mephistopheles as that being than which nothing greater in evilness can be conceived. Although the existence of the Devil would undoubtedly be a greater evil than the idea of the Devil, this definition fortunately does not establish the reality of such an unwelcome being.

René Descartes also proposed an ontological proof of the existence of God. He wrote in a letter, "I dare to boast that I have found a proof of the existence of God which I find fully satisfactory and by which I know that God exists more certainly than I know the truth of any geometrical proposition." His argument, in brief, is that God, who is perfect, must exist because existence is an element of perfection. A perfect supreme being cannot be merely an imaginary being.

In his *Third Meditation* Descartes wrote, "I shall now close my eyes, stop my ears, turn away all my senses, even efface from my thoughts all images of corporeal things, or at least, because this can hardly be done, I shall consider them as being vain and false." By introspection he found, "I am a thing which thinks, that is to say, which doubts, affirms, denies, knows a few things, is ignorant of many, which loves, hates, wills, does not will, which also imagines, and which perceives." By self-contemplation Descartes had already concluded that his own existence was beyond all doubt: "I think, therefore I am."

In the *Third Meditation*, he continued,

> There remains then only the idea of God, in which I must consider whether there is anything which could not have come from me. By the name of God I understand an infinite substance, eternal, immutable, independent, omniscient, omnipotent, and by which I and all other things which exist (if it be true that any such exist) have been created and produced. But these attributes are so great and eminent, that the more attentively I consider them, the less I am persuaded that the idea I have of them can originate in me alone. And consequently I must necessarily conclude from all I have said hitherto, that God exists.

He perceived himself as an imperfect person who nonetheless had glimpses of perfection. Whence came these revelations of perfection? Not from himself, nor any other imperfect being; hence they must have come from God. Starting with the fact that he existed, Descartes came to the conclusion that God's existence was equally beyond doubt.

Descartes then cast his net more widely and found that he himself possessed a soul as a result of God's existence. Other human beings, he conceded, must also have souls. But not animals, who had no souls and were placed on Earth for the benefit of mankind. When faith intrudes, staining the purity of natural theology, the conclusions drawn in the name of reason tend to be whatever the inquirer desires.

Immanuel Kant rejected the Cartesian argument on essentially the grounds used by Aquinas against Anselm: "The concept of a supreme being is in many respects a very useful idea, but just because it is a mere idea, it is incapable alone by itself of enlarging our knowledge on what exists. It is not even competent to enlighten us as to the possibility of any existence beyond that which is known in and through experience." The force of the ontological argument springs from religious preconceptions, and as an intellectual exercise in pure reason there is little doubt the conclusions reached are unwarranted. Agnostics are unconvinced and the argument fails in its main purpose.

* * *

The moral argument – especially favored by Kant – takes for granted the premise that ethical principles and moral standards are the province of religion, thereby leading to the conclusion that God is the source of all distinction between right and wrong, between good and evil, that without God there can be no perfect goodness. The argument is not accepted by agnostics, who nowadays regard the social evolution of moral codes as more natural.

The scorpion's sting that paralyzes the argument is the abundance of evil in the world to which religious institutions and their members have in the past made disproportionate contributions.

David Hume in his essay *The Immortality of the Soul* poured scorn on the argument: "Let us consider the moral arguments, chiefly those derived from the justice of God, who is supposed to be interested in the future punishment of the vicious and the reward of the virtuous.

But these arguments are grounded on the supposition that God has attributes beyond what he has exercised in the universe with which we are acquainted." A century later, in *The Utility of Religion*, John Stuart Mill said that in no way "can the government of nature be made to resemble the work of a being at once good and omnipotent." If the supreme being is accountable for all this wretchedness, then that being is either not perfect goodness or not all-powerful. Either evil does not exist (contrary to the daily news), or it exists and therefore God cannot be both all-good and all-powerful. Bertrand Russell in *Why I Am Not a Christian* rams the point home when he says, we "could take the line that some of the agnostics took – a line which I often thought was a very plausible one that as a matter of fact this world that we know was made by the devil at a moment when God was not looking."

Plausibly the moral codes by which we live and the ethical sparks that illumine our lives are as old as *Homo sapiens*. During the tens and hundreds of millennia of human prehistory, those social groups not consisting of mutually supporting individuals had little chance of surviving. Ancient naturally selected codes of social behavior became eventually the ordinances decreed by gods. The gods legislated the laws, rewarded those who obeyed, and punished those who disobeyed. Now the gods have fled from the world taking with them the seals of authenticity affixed to our moral standards. The ancient codes that distinguished between right and wrong, between good and evil, are now myths, the objects of derision, irrelevant in political and legal deliberations. We are taught no one is evil; the cause is hereditary, or environmental, or the result of mental illness, and the fault is never our own. All is not lost, however, for the codes of behavior of primitive men and women linger on, and deep within us we know what is right and wrong, what is good and evil. The codes are still active, preserving society, and remain our primary defense in the survival game.

The moral proof of the existence of God, once the most persuasive of all arguments, has become the least convincing in an age that takes for granted that the physical universe is basically without intrinsic ethical content.

*   *   *

The cosmological argument seeks to establish proof by showing that the universe is neither self-creating, self-sustaining, nor self-sufficing, and to show that the existence of God repairs one or more of these deficiencies. The closer we identify ourselves with the belief systems of the ancient world the more compelling becomes the cosmological argument. The argument nowadays relates to matters primarily scientific and not theological. Thus, the discovery that space and time are physical and therefore created with the universe has greatly affected the nature of cosmogenesis.

In this book all variants of the cosmological proof are off-limits. I hold that it is impossible to find concrete proof of the existence of God in the framework of any universe, for all universes are devised and figured by the human mind.

John Laird in his book *Theism and Cosmology* expresses similar views: "One of the principal obstacles that beset all arguments from the world to God is the doubtful legitimacy of arguing from the relations or connections within the cosmos to a similar relation or connection between the cosmos and some transcendental being." We may relate God and the Universe, both of which are unknown, but it is unfitting to relate God with a particular universe.

*   *   *

The teleological or design argument seeks to demonstrate that the universe is designed to serve a purpose and meet a certain end. Some of the conditions determining the fitness of a universe for inhabitation by life were considered in the previous chapter. Immanuel Kant, who took great interest in the proofs of God's existence, said the argument from design is "the oldest, the clearest, and the most in conformity with the common reason of humanity." But the design argument now lacks conviction and is in retreat. What once seemed intentional design has become implicit in the basic makeup of the universe, a makeup that many hope to show is essential to the existence of the universe. Thus, the design argument becomes a variation of the cosmological argument.

*   *   *

Anselm's definition of God as "that being than which nothing greater can be conceived" prompts the following train of thought. We consider two mutually exclusive sets: the first set consists of all conceivable things, whether imaginary or real, and the second set consists of all inconceivable things. Thus, general relativity, quantum mechanics, quarks, little green men, and the cow that jumped over the moon are members of the first set. Presumably members of the second set exist but are beyond our conception.

In which set do we place God? If in the first set, amidst conceivable things, then by Anselm's definition God exists at the limit of greatness. In that case reason alone cannot assure us that greater things are not members of the second set. But if in the second set, amidst inconceivable things, we are denied any means of measuring greatness. Therefore we must modify Anselm's definition into something like: "God is all and inconceivable." God is all, necessarily, for otherwise we would have no assurance that other things of an inconceivable nature might be greater.

Anselm's definition makes us realize that we are discussing what can and cannot be conceived by the human mind. God is diminished when brought into the first set and made comprehensible to human beings. As Aquinas said, the divinity "exceeds by its immensity every form that our intellect attains." We are compelled as natural theologians to place God in the second set amidst all that is inconceivable, leaving us, unfortunately, with no more than the empty name of an ungraspable entity of problematical existence.

*   *   *

From mythology we inherit the custom of referring to God as a personal being – as He or She – as the Father or Mother – endowed with superhuman characteristics. In order to have an intimate relationship of the kind human beings once had with the spirits and godlings of long ago, God is personified and brought into the first set, the set of conceivable things. It is then possible, as in the medieval and other monotheistic versions of the mythic universe, to compare God with other conceivable things and say that God is greater than all the rest. But making God into a personal being diminishes the nature of God.

Anselm's definition inspires us to define the Universe as "that thing than which nothing greater can be conceived." This definition, however, is not good enough. The Universe lies in the second set, the set of inconceivable things, and exceeds anything that can be conceived. A better definition is: "the Universe is all-inclusive and inconceivable."

When the Universe and God are both brought into the first set (the set of conceivable things) we declare that one is personal and the other impersonal. When both are left in the second set (the set of inconceivable things), where they rightly belong, it is impossible to make any distinction between the two. Pure reason by itself informs us that we must leave God in the second set without any identifying features, anthropomorphic or otherwise, having the same definition as the Universe.

*   *   *

The myriad gods are models of God in much the same way as the myriad universes are models of the Universe. The universes are the masks of the Universe, and the gods are "the masks of God" (to borrow a phrase used by the anthropologist Joseph Campbell). Even a personal and loving supreme being is a mask, a god, conceived and figured by the human mind. The gods and universes are grand unifying concepts occupying the set of conceivable things. Both God and the Universe, on the other hand, are beyond understanding and occupy the set of inconceivable things.

We may arrange our ideas in such a way that a particular god is the creator of a particular universe. But we are not at liberty to do the same with God and the Universe because both are beyond understanding. Furthermore, it would be unfitting to adopt the halfway argument that God is the creator of a universe – the one currently in vogue – because universes are devised by the human mind. It would be equally unfitting to argue that a particular god – the one currently in vogue – is the creator of the Universe, because gods also are devised by the human mind.

The difficulty with most arguments in natural theology is that God and gods, and Universe and universes are never properly distinguished. Thus, a not uncommon statement is that "the cosmos does not have in itself a sufficient reason for its own existence." This statement is clearly about a universe. It is then argued that the existence of a particular god remedies the deficiency. It is always within the power of human wit to show that a god is greater than a cosmos, that the existence of the latter depends on the former, and hence the former must exist.

When we realize it is the existence of the Universe that must be explained, all such arguments fail. Those who call on their gods to explain their universes must, by symmetry anticipate others of a different conviction who call with equal validity on their universes to explain their gods.

We may relate gods and universes (always stressing that both are models), or relate God and the Universe (always stressing that both are unknown and inconceivable), but we cannot relate a god and the Universe or relate a universe and God. We cannot enter the set of inconceivable things and hold discourse on what lies beyond our comprehension. The terminology of eternal, infinite, omnipotent, and omniscient, although impressive, has meaning only in the context of models that lie in the set of conceivable things.

*   *   *

Both God and the Universe are defined as all-inclusive and inconceivable. But this creates a redundancy of all-inclusive things of an inconceivable nature. The definition contains nothing to prevent us from supposing that both are plausibly one and the same thing. Such a startling hypothesis has the merit of conceptual economy and the advantage of unifying things beyond the limit of human understanding.

God and Universe are one and the same without distinction. "God is all" takes on the wider meaning of "all is God." God and the Universe unify into UniGod.

Let me hasten to say that equating God and Universe is not another version of pantheism. Pantheism is the belief that gods in the form of great nature spirits are immanent within but not transcendent over the universe. A pantheistic world view lies between the old magic and mythic universes, and where it lies on the keyboard, either toward the magic end or the mythic end, depends on the extent of the unifying power ascribed to the immanent spirit gods.

Theological and cosmological traditions in Western society prohibit the equating of gods and universes, as the Dutch philosopher Baruch Spinoza found in the seventeenth century. In our theologies the favored god is confused with God; in our cosmologies the favored universe is confused with the Universe; and our concepts of gods on the one hand and of universes on the other have diverged to the stage where they apply to totally dissimilar realms of concepts. We dare not tamper with them and we naturally shudder at the thought of equating models of God to models of the Universe.

Rejection of the possibility of a God–Universe, or UniGod, perhaps explains why we find ourselves in need of proofs of God's existence. Human beings have abstracted from nature all its holiness and ascribed it to the gods, leaving the natural world dead and soulless. The gods have fled into their surrealistic worlds of improbable existence, taking away from the world all that we call divine. We ourselves have transformed God into a fiction that cannot be proved true.

Who doubts the existence of the unknown and unknowable Universe of which we are a part? The history of cosmology reveals numerous universes, and when we extrapolate from the past to the future, we think it not unreasonable to suppose that a great many universes will exist in the tens, hundreds, and thousands of millennia to come. And we must not forget the universes devised by extraterrestrial intelligent beings. Each universe masks the Universe whose reality, of which we are a part or an aspect, is beyond all doubt.

Given that the Universe and God are one and the same – the Cloud of Unknowing – we cannot doubt the existence of God, for the existence of the Universe is beyond doubt. This is surely the ultimate ontological proof of the existence of God. When pressed to its limit, the ontological argument that God is all-inclusive and inconceivable leads to no other conclusion. If we recognize that God and Universe are interchangeable names referring to the all-inclusive and inconceivable, then the reality of God is beyond doubt. This proof of the existence of God springs from agnostic soil.

*   *   *

Cosmology and theology are linked by six cardinal dualities. The first and second are the two commensurate dualities: God and Universe, and gods and universes. The third and fourth are the two incommensurate dualities: God and gods, and Universe and universes. The fifth and sixth are the two incompatible dualities: God and universes, and Universe and gods.

First duality – God and Universe: We have argued they are one and the same thing. Cosmology and theology thereby recover an original partnership in an enterprise forever seeking to unmask ultimate reality. The argument has the merit of theoretical economy. As Newton said, "Nature is pleased with simplicity and affects not the pomp of superfluous causes."

Second duality – gods and universes: On one hand we have the universes, and on the other the gods spinning their cosmotheistic fabrics. The universes are models of the Universe and the gods are models of God. They all lie within the set of conceivable things. In pantheism both are woven together into a unigod (one of many models of the UniGod). Johann Goethe, poet and sage, wrote:

Nature! We are surrounded and embraced by her: powerless to separate ourselves from her, and powerless to penetrate beyond her. . . . She has always thought and always thinks; though not as a man, but as Nature. She broods over an all-comprehending idea, which no searching can find out. . . . She has neither language nor discourse; but she creates tongues and hearts, by which she feels and speaks. . . . She is all things.

This is one of the finest examples of a unigod serving as a model of the UniGod. All is well provided we resist the temptation to treat any one model as the final revelation. Those religious persons claiming to know what the Universe is because God has told them are imprisoned within their out-of-date models.

Third duality – God and gods: Polytheistic and monotheistic masks of God are commonly venerated as the true face. Mistaking the model for the thing itself is as rife today in the world's vast populations as at any time in the past, and is the source of confused thinking in many of the arguments that attempt to prove the existence of God.

Fourth duality – Universe and universes: A duality as old as cosmology that is the theme of this book. Wherever we alight in the history of cosmology we find the current universe mistaken for the Universe. This misidentification is as rife today as at any time in the past.

Fifth duality – God and universes: This duality is the source of confused thinking in the cosmological and teleological arguments proving the existence of God. On one side are the universes conceivable and inglorious, and on the other side a vision of God inconceivable and glorious. Theology uses these incompatible premises as a springboard for demonstrating the necessity of a supreme being. A feeling of urgency lies behind these demonstrations. The glory has been extracted from the ambient world and given to a supreme being, and without the demonstrable existence of that being we are left with only an inglorious residue.

Sixth duality – Universe and gods: This duality is a possible source of much confusion in atheism. The gods are seen as hangovers from mythology and are discarded as incompatible with a vision of the Universe. But the cosmic vision, like the gods, is no more than a model.

*   *   *

We feel a strong urge to believe in God, and this desire derives not from arguments of pure reason. The urge is emotional and admittedly irrational in the context of the atomic and neurological structures of the modern physical universe. Rationalists resist the irrational urge on the grounds that it emerges from the jungle of our cultural heritage. Many rationalists and most agnostics realize, however, that the absence of proof of God is not proof of God's absence, and take an occasional interest in arguments claiming to show that we have the necessary and/or sufficient proof of the existence of a supreme being.

One might legitimately argue that as a result of the atheistic rejection of the gods, our views of reality are pallid and inane, and our views of life are devoid of metaphysical meaning. Our cultural heritage impels us to believe in God, or something similarly mysterious and all-inclusive, for long ago it stole from the phenomenal world the elements essential for a life of personal meaning, and gave those elements to the gods whose function it is to share them with us.

We have only to imagine the home having its own house god or hearth goddess, who emanates an ambiance of warmth and friendliness and wards off danger, who is acknowledged by libations and flowers, for the home to seem secure and restful. This fanciful illustration shows how deep within us springs an urge to live in fellowship with the gods.

In *The First Three Minutes* (1970), Steven Weinberg made the remark, "The more the universe seems comprehensible, the more it seems pointless." The loss of the medieval universe with its unification of human experience, followed by the rise of the monolithic mechanistic universe, has left us fluttering aimlessly like moths among the mechanisms, lamenting how pointless it all seems.

Belief in a mysterious God that is all-inclusive and inconceivable enables us to view with equanimity the prospect that our universes are not the Universe and never have been nor ever will be. Belief in an unknown and unknowable God, or Universe, or UniGod, counsels humility and hope, not arrogance and despair.

If we can think that all is far from known and God is perhaps the Universe, then without further intellectual commitment we avoid the dreariness of atheism and the emptiness of agnosticism. By equating God and the Universe we give back to the world what long ago was taken from it. The world we live in with our thoughts, passions, delights, and whatever stirs the mortal frame must surely take on a deeper, richer meaning. Songs become more than longitudinal sound waves, sunsets more than atmospheric scattering of transverse electromagnetic oscillations, inspirations more than the discharge of neurons, all touched with a mystery that deepens the more we contemplate and seek to understand.

Alfred Whitehead wrote, "the theme of cosmology is the basis of religion." The converse statement, the theme of religion is the basis of cosmology, rings with equal truth. This perhaps explains why interest in cosmology grows when commitment to religion declines. Both have the same basic theme with the one often complementing the other. "Science without religion is lame, religion without science is blind," said Albert Einstein.

*   *   *

On occasion in my school days, Canon Morrow sauntered across the cathedral close and took over a scripture lesson in one of the classrooms. In front of the class, in a ringing Church-of-England voice and with dramatic gestures, he brought the past to life. Acting the part of a Roman centurion, he would leap aside from Queen Boadicea's chariot as it swept by on the field of battle, and with arm upraised, hurl his imaginary javelin at a barbarian chieftain, and skipping across the room to act the part of the chieftain, he would turn and stagger forward as the javelin pierced his breast. I sat riveted in my seat as one of the enrapt pupils. After the canon had left to pursue his other cathedral duties, our teacher would resume control, sometimes making guarded comments that have since led me to realize he was an agnostic. By modern standards, education at a cathedral school many years ago in England was totally mistaken and grossly misdirected. But it gave me dragons to slay and set me on paths of inquiry that have lasted a lifetime. I thank the gods I escaped the inanities of modern education.

Do not deny the gods. Fight them if you will. But grovel, and in their contempt they will scoop you into the holy mincing machine of incarnadine wars. Hate them! Curse them! Though they may crush you, they will not despise you. But if you ignore them, then beware! For in their anger they will inflict on you nameless horrors of body and mind. Only fools deny the hereditary gods that live within us.

19 Learned Ignorance

As the light of knowledge spreads and brightens,
the shadows of learned ignorance gather and darken

In his work *On Learned Ignorance*, written in 1440, Cardinal Nicholas of Cusa argued that although the darkness of "unlearned ignorance" disperses in the light of growing knowledge, there is another side to ignorance, which he called "learned ignorance," that grows with knowledge. "No man, not even the most learned in his discipline, can progress farther along the road to perfection than the point where he is found most knowing in the very ignorance that embraces him; and he will be the more learned the more he comes to know himself for ignorant."

Consider the unlearned person, unaware of his ignorance, who thinks he knows everything! As knowledge increases, ignorance decreases, yet this kind of ignorance – unlearned ignorance – is merely the absence of knowledge. With knowledge comes an awareness of ignorance – learned ignorance – and the more a person knows, the more aware that person becomes of what he does not know. Learned and unlearned ignorance are like day and night.

The principle of learned ignorance at first comes as a surprise. "Knowledge is power" says the proverb. We acquire learning seeking to extend the horizon of our knowledge. Education uplifts the mind and dispels ignorance. Issues that arise in the learning process that at first are puzzling are subsequently resolved in the corpus of greater knowledge. But, as the learned cardinal said, the penalty of knowledge is doubt.

Life begins full of confidence and ends full of doubt. As knowledge grows, new facts and fresh ideas cast shadows of uncertainty over old facts and ideas. Previous knowledge must repeatedly be revised and reinterpreted. In time the new knowledge inevitably reaps the harvest of further doubt. Uncertainty becomes one's constant companion.

Learned ignorance – awareness of ignorance – like entropy seems never to decrease but always increase. It urges us to seek certainty by acquiring greater knowledge, which when attained, unfailingly creates further uncertainty. Solve one problem and you create many more.

Tentatively, I suggest that the first law of knowledge is *the conservation of ignorance*. (The first law of thermodynamics is the conservation of energy.) As unlearned ignorance decreases, learned ignorance increases, and their sum stays more or less constant. The more we know the more aware we become of what we do not know.

As with individuals so with societies. Consider a society blissfully unaware of its ignorance. With confidence its members declare they know everything worth knowing. What is not known – the mysteries – lies in the safe-keeping of their gods. This condition is eventually disturbed by the rise of novel ideas and the discovery of awkward facts. As the universe develops, so learned ignorance grows. A universe is never complete, never perfect, and the more that is known, the more apparent become its imperfections. As the light of knowledge spreads and brightens, the shadows of learned ignorance gather and darken.

*   *   *

I suggest for the second law of knowledge: *the ratio of learned ignorance to knowledge tends always to increase.* (In thermodynamics the second law states that entropy in an isolated system tends always to increase.) There is no way of measuring learned and unlearned ignorance and knowledge, and the first and second laws must be regarded as qualitative relations of a subjective nature. Learned ignorance – a conscious awareness of ignorance – is one of the main agents causing universes to evolve. It increases at least as fast as knowledge. For if it increased slower, in time what is known to be unexplained would diminish in proportion to what is explained, and as a result societies would progressively become more secure and their universes more durable. History shows that the opposite happens: magicomythic universes endure for long periods of time, whereas more complex mechanistic universes endure for relatively short periods.

It seems safe to say that learned ignorance grows faster than knowledge, or at least as fast, otherwise the advance of knowledge would slow down and finally come to a halt, and universes would reach an end state and cease to evolve. But history shows that the pace quickens, not slackens.

The greater the knowledge the greater the doubt. Consequently, the greater the urge to banish doubt with more knowledge. The rock that Sisyphus must forever push up the mountain, like learned ignorance, grows ever heavier. People in the future with their vast knowledge will bear burdens of learned ignorance immensely heavier than our own.

*   *   *

The history of science teaches us that knowledge leads to doubt, and doubt in turn spurs the search for further knowledge, leading to further doubt. Why do we bother with the pursuit of knowledge and its reward of doubt, why not stay happily unlearned, untroubled by uncertainty? The answer seems to lie partly in the belief that the end to the search for knowledge is close at hand. The winning-post to final knowledge is within reach.

Soon, we are told, we shall have no more diseases and no more poverty. Theories that purport to "explain everything" are in the news. The Sisyphean struggle is at an end and the rock is about to reach the top of the mountain. In our laboratories, observatories, and temples we are finally unmasking the face of the Universe.

Can knowledge ever be complete and perfect? Can it advance to the point where a person can legitimately claim to know everything worth knowing and there is nothing more to understand? Can unlearned ignorance ever be totally zero? I do not think so. Maybe there is a third law of knowledge: *unlearned ignorance can never be reduced to zero.* (The third law of thermodynamics states that the temperature of a system can never be reduced to zero.) Knowledge never attains the point where it is complete and there is nothing further to learn.

*　　*　　*

A mathematical system is constructed from a basic set of self-consistent postulates (or axioms). In terms of this system a mathematician determines if certain mathematical propositions are true or false. In 1910, 1912, and 1913 Bertrand Russell and Alfred North Whitehead published the three volumes of *Principia Mathematica*. In this monumental treatise they sought to construct a universal and complete system of mathematical logic such that every mathematical statement could be shown to be either true or false.

In 1931 Kurt Gödel, a young Austrian mathematician, published a short paper: "Formally undecidable propositions of *Principia Mathematica* and related systems," and dashed all hopes of ever realizing a complete and self-consistent mathematical system. Gödel's incompleteness theorem showed that to any system of mathematical logic there are propositions that can be neither proved nor disproved. When the system is appropriately enlarged, previous undecidable propositions can then be shown to be true or false. There is a price to pay. For this enlarged system new undecidable propositions now exist. Gödel's theorem demonstrates that in any system of mathematical logic it is possible to formulate propositions that are decidable only in the context of larger systems.

Language, though less rigorous than mathematics, provides illustrations of Gödel's theorem. Thus, in a town, the one male barber shaves only all men who do not shave themselves. Who shaves the barber? The question becomes answerable by enlarging the population of the town to include females. A possible answer then could be, "his wife." "This statement is false" exemplifies an inconsistent system: if true, it is false; if false, it is true. Gödel pointed out the analogy between his incompleteness theorem and the paradox of the liar. Epimenides declared "all Cretans are liars," yet was himself a Cretan. If he told the truth he lied because he was a Cretan; only if he lied could he have told the truth.

Gödel's incompleteness theorem awakens old philosophical problems. Can we justify reason by using only the methods of reason? The answer is no! This skeleton, briefly exposed to view by David Hume in the eighteenth century, is traditionally kept locked in the closet.

$*$ $*$ $*$

Gödel's incompleteness theorem is analogous to Nicholas of Cusa's principle of learned ignorance. Leaned ignorance is an awareness of incomplete knowledge. The more we enlarge knowledge the more aware we become of our ignorance. We only think we know when unaware that we do not know. In all systems of logic there are undecidable propositions, and the larger the system, the larger their number. In all rational systems of knowledge there is an awareness of incompleteness, and the larger the system, the greater the awareness.

Beyond all systems stands the Universe in a cloud of unknowing, and "he will be the more learned, the more he comes to know himself for ignorant."

Bibliography

Further sources and readings may be found in the *Dictionary of Scientific Biography* (Scribner's, New York, 1972), the *Encyclopedia of Philosophy* (Macmillan, New York, 1967), and the 11th edition of the *Encyclopaedia Britannica* (Cambridge University Press, 1910).

E. A. Abbott, *Flatland*. Dover, New York, 1992.

R. E. Allen, *Greek Philosophy: Thales to Aristotle*. Free Press, New York, 1966.

S. Angus, *The Mystery-Religions: A Study in the Religious Background of Early Christianity*. Dover, New York, 1975.

E. Anscombe and P. T. Geach, editors, *Descartes: Philosophical Writings*. Nelson, Edinburgh, 1954.

S. Arrhenius, *Worlds in the Making*. Harper, London, 1908.

I. Asimov, *The Gods Themselves*. Panther, New York, 1974.

Augustine, *Confessions*, translator R. S. Pine-Coffin. Penguin, Harmondsworth, Middlesex, UK, 1961.

M. Aurelius, *Meditations*, translator M. Staniforth. Penguin, Harmondsworth, Middlesex, UK, 1964.

W. C. Bark, *Origins of the Medieval World*. Stanford University Press, Stanford, 1958.

J. D. Barrow, *Theories of Everything*. Clarendon Press, Oxford, 1991.

J. D. Barrow and F. J. Tipler, *The Anthropic Cosmological Principle*. Oxford University Press, Oxford, 1986.

C. L. Becker, *The Heavenly City of the Eighteenth-century Philosophers*. Yale University Press, New Haven, 1968.

R. Berendzen, R. Hart, and D. Seeley, *Man Discovers the Galaxies*. Science History Publications, New York, 1976.

G. Berkeley, *A Treatise Concerning the Principles of Human Knowledge*, editor C. M. Turbayne. The Library of Liberal Arts, Indianapolis, 1957.

M. Berman, *The Reenchantment of the World*. Cornell University Press, Ithaca, 1982.

C. Blakemore, *Mechanics of the Mind*. Cambridge University Press, New York, 1977.

N. Block, F. Owen, and G. Güzeldere, editors, *The Nature of Consciousness: Philosophical Debates*. MIT Press, Cambridge, Massachusetts, 1997.

Boethius, *The Consolation of Philosophy*, translator V. E. Watts. Penguin, Harmondsworth, Middlesex, UK, 1969.

D. Bohm, *Quantum Theory*. Prentice-Hall, Englewood Cliffs, New Jersey, 1951.

H. Bondi, *Cosmology*. Cambridge University Press, Cambridge, 1960.

J. L. Borges, "The garden of forking paths," in *Labyrinths: Selected Stories and Other Writings*, editors D. A. Yates and J. E Irby. New Directions, Norfolk, Connecticut, 1962.

A. M. Bork, "The fourth dimension in nineteenth-century physics." *Isis*, **55**, 326, 1964.

M. Born, *Natural Philosophy of Cause and Chance*. Clarendon Press, Oxford, 1949.

J. Brown, *Minds, Machines and the Multiverse*. Simon and Schuster, New York, 1999.

G. Bruno, *Gesammelte Werke*, editor L. Kuhlenbeck. Eugen Diederichs, Jena, 1904.

J. Buchler, *The Philosophical Writings of Peirce*. Dover, New York, 1955.

J. D. Burchfield, *Lord Kelvin and the Age of the Earth*. Science History Publications, New York, 1975.

J. Burnet, *Early Greek Philosophy*. Meridian Books, New York, 1957.

E. A. Burtt, *The Metaphysical Foundations of Modern Physical Science*. Doubleday, Garden City, New York, 1932.

J. B. Bury, *History of the Later Roman Empire*. Dover Publications, New York, 1958.

V. Bush, *Science Is Not Enough*. Morrow, New York, 1967.

J. Campbell, *The Hero with a Thousand Faces*. Princeton University Press, Princeton, 1968.

F. Capra, *The Turning Point: Science, Society and the Rising Culture*. Simon and Schuster, New York, 1982.

B. Carter, "Large-number coincidences and the anthropic principle," in *Confrontations of Cosmological Theories with Observational Data*, editor M. S. Longair. Reidel, Dordrecht-Holland, 1974.

R. Cavendish, *Visions of Heaven and Hell*. Harmony Books, New York, 1977.

B. J. Chalmers, "The puzzle of conscious experience." *Scientific American*, December 1995.

D. J. Chalmers, *Conscious Mind: In Search of a Fundamental Theory*. Oxford University Press, Oxford, 1996.

D. J. Chalmers, *The Conscious Mind*. Oxford University Press, New York, 1996.

M. Chown, *The Universe Next Door*. Headline, London, 2001.

F. Close, *The Cosmic Onion*. Springer-Verlag, New York, 1995.

I. B. Cohen, *The Newtonian Revolution*. Cambridge University Press, New York, 1981.

J. E. Cohen, *How Many People Can the Earth Support?* Norton, New York, 1995.

N. Cohn, *Europe's Inner Demons: An Enquiry Inspired by the Great Witch Hunt*. Meridian Books, New American Library, New York, 1977.

F. M. Cornford, *Plato's Cosmology*. Routledge and Kegan Paul, London, 1937.

F. M. Cornford, *From Religion to Philosophy*. Harper and Row, New York, 1957.

F. H. C. Crick, "Thinking about the brain." *Scientific American*, September 1979.

F. H. C. Crick, *Life Itself: Its Origin and Nature*. Simon and Schuster, New York, 1982.

A. C. Crombie. *Augustine to Galileo*. Vol. 1, *Science in the Middle Ages 5th–13th Centuries*; Vol. 2, *Science in the Later Middle Ages and Early Modern Times 13th–17th Centuries*. Mercury Books, London, 1964.

K. Crosswell, *The Universe at Midnight: Observations Illuminating the Cosmos*. Free Press, New York, 2001.

F. Cumont, *The Mysteries of Mithra*. Dover Publications, New York, 1956.

F. Cumont, *Astrology and Religion Among the Greeks and Romans*. Dover Publications, New York, 1960.

Dante Alighiere, *The Divine Comedy: Paradise*, translators D. Sayers and B. Reynolds. Penguin, Harmondsworth, Middlesex, UK, 1962.

C. Darwin, *The Origin of Species by Means of Natural Selection, or the Preservation of Favoured Races in the Struggle for Life*. Penguin Books, Harmondsworth, Middlesex, UK, 1968. First published by John Murray in 1859.

P. C. W. Davies, *The Runaway Universe*. Dent, London, 1978.

P. C. W. Davies, *Other Worlds*. Simon and Schuster, New York, 1980.

P. C. W. Davies, *The Accidental Universe*. Cambridge University Press, Cambridge, 1982.

P. C. W. Davies and J. Brown, editors, *Superstrings: A Theory of Everything*. Cambridge University Press, Cambridge, 1988.

R. Descartes, *Discourse on Method and the Meditations*, translator F. E. Sutcliffe. Penguin, Harmondsworth, Middlesex, UK, 1968.

D. Deutsch, *The Fabric of Reality*. Penguin, Harmondsworth, Middlesex, UK, 1997.

B. S. DeWitt, "Quantum mechanics and reality." *Physics Today*, September 1970.

B. S. DeWitt and N. Graham, *The Many-worlds Interpretation of Quantum Mechanics*. Princeton University Press, Princeton, 1973.

Diogenes Laertius, *Lives of Eminent Philosophers*, translator R. D. Hicks, 2 volumes. Loeb Classical Library, New York, 1935.

P. A. M. Dirac, "The evolution of the physicist's picture of nature." *Scientific American*, May 1963.

S. Drake, *Galileo at Work*. University of Chicago Press, Chicago, 1978.

R. Dubos, *The Dreams of Reason*. Columbia University Press, New York, 1961.

R. S. Dunn, *The Age of Religious Wars, 1559–1715*. Norton, New York, 1979.

J. W. Dunne, *An Experiment with Time*. Macmillan, New York, 1927.

E. Durkheim, *Elementary Forms of the Religious Life*. Free Press, New York, 1965.

F. J. Dyson, "The future of physics." *Physics Today*, September 1970.

A. S. Eddington, *Space, Time, and Gravitation*. Cambridge University Press, 1920. Reprinted by Harper Torchbooks, New York, 1959.

A. S. Eddington, *The Expanding Universe*. Cambridge University Press, Cambridge, 1933. Reprinted by the University of Michigan Press, Ann Arbor, 1958.

A. Einstein, *The World as I See It*. The Philosophical Library, New York, 1934.

L. Eiseley, *The Immense Journey*. Vintage Books, New York, 1957.

M. Eliade, *The Myth of the Eternal Return*. Princeton University Press, Princeton, 1971.

M. Eliade, *A History of Religious Ideas: From the Stone Age to the Eleusian Mysteries*. University of Chicago Press, Chicago, 1978.

A. F. Elkin, *The Australian Aborigines*. The Natural History Library, Doubleday, Garden City, New York, 1964.

G. F. R. Ellis, "The world's environment: The universe." *South African Journal of Science*, **75**, 529, 1979.

G. F. R. Ellis and G. B. Brundrit, "Life in the infinite universe." *Quarterly Journal of the Royal Astronomical Society*, **20**, 37, 1974.

S. Epstein, "The self-concept revisited." *American Psychologist*, **28**, 404, 1973.

A. Erman, *The Ancient Egyptians: A Sourcebook of their Writings*, translator A. M. Blackman. Harper and Row, New York, 1966.

E. E. L'Estrange, *Witch Hunting and Trials*, Routledge and Kegan Paul, London, 1929.

J. Evans, *The History and Practice of Ancient Astronomy*. Oxford University Press, New York, 1998.

E. Evans-Pritchard, *Theories of Primitive Religion*. Oxford University Press, London, 1965.

E. Evans-Pritchard, *A History of Anthropological Thought*. Basic Books, New York, 1981.

P. Farb, "How do I know you meant what you mean?" *Horizon*, Autumn 1968.

B. Farrington, *Greek Science, Its Meaning for Us*. Vol. 1, *Thales to Aristotle*. Vol. 2, *Theophrastus to Galen*. Penguin, Harmondsworth, Middlesex, UK, 1944.

G. Feinberg, *What Is the World Made of? Atoms, Leptons, Quarks, and Other Tantalizing Particles*, Doubleday, Garden City, New York, 1977.

T. Ferris, *The Whole Shebang: A State of the Universe(s) Report*. Simon and Schuster, New York, 1997.

R. Feynman, *The Character of Physical Laws*. M.I.T. Press, Cambridge MA, 1967.

A. Flew, editor, *Body, Mind, and Death*. Macmillan, New York, 1964.

P. Frank, *Einstein: His Life and Times*. Knopf, New York, 1947.

H. Frankfort, G. A. Frankfort, J. A. Wilson, and T. Jacobsen, *Before Philosophy: The Intellectual Adventure of Ancient Man*. Penguin, Harmondsworth, Middlesex, UK, 1949.

J. T. Fraser, editor, *The Voices of Time*. University of Massachusetts Press, Ambits, 1981.

J. G. Frazer, *The Golden Bough: A Study in Magic and Religion*, abridged edition. Macmillan, New York, 1945.

A. Freemantle, *The Age of Belief*. New American Library, New York, 1954.

E. Fromm and R. Xirau, editors, *The Nature of Man*. Macmillan, New York, 1968.

R. M. Gale, editor, *The Philosophy of Time: A Collection of Essays*. Humanities Press, Atlantic Highlands, NJ, 1968.

P. L. Galison, "Minkowski's space-time." *Historical Studies in the Physical Sciences*, **10**, 85, 1979.

G. Gamow, *The Creation of the Universe*. Viking, New York, 1952.

M. Gardner, *Fads and Fallacies in the Name of Science*. Dover, New York, 1957.

M. Gardner, *The Ambidextrous Universe*. Basic Books, New York, 1964.

H. Genz, *Nothingness: The Science of Empty Space*. Perseus Books, Cambridge, Massachusetts, 1999.

J. Gimpel, *The Medieval Machine: The Industrial Revolution of the Middle Ages*. Penguin Books, New York, 1977.

O. Gingerich, *The Eye of Heaven*. Springer-Verlag, New York, 1995.

T. Gold, editor, *The Nature of Time*. Cornell University Press, Ithaca, 1967.

K. Gödel, *On Formally Undecidable Propositions*. Basic Books, New York, 1962.

P. H. Gosse, *Omphalos: An Attempt to Untie the Geological Knot*. J. van Voorst, London, 1857. (Ox Bow Press, 1998).

J. R. Gott III, *Time Travel in Einstein's Universe*. Houghton Mifflin, New York, 2001.

S. J. Gould, *Wonderful Life*. Norton, New York, 1989.

E. Grant, *Physical Science in the Middle Ages*. Cambridge University Press, Cambridge, 1977.

E. Grant, *Much Ado About Nothing*. Cambridge University Press, Cambridge, 1981.

B. Greene, *The Elegant Universe: Superstrings, Hidden Dimensions, and the Quest for the Ultimate Theory*. Norton, New York, 1999.

J. C. Greene, *Darwin and the Modern World View*. Louisiana State University Press, Baton Rouge, 1961.

J. C. Greene, "Evolution and progress." *Johns Hopkins Magazine*, October 1962.

R. L. Gregory, *Mind in Science: A History of Explanations in Psychology and Physics*. Cambridge University Press, New York, 1981.

J. Gribbin and M. J. Rees, *Cosmic Coincidences*. Black Swan, London, 1991.

C. G. Gross, *Brain, Vision, Memory: Tales in the History of Neuroscience*. MIT Press, Boston, 1998.

A. Guillaume, *Islam*. Penguin, Harmondsworth, Middlesex, UK, 1954.

A. H. Guth, *The Inflationary Universe*. Perseus Press, New York, 1997.

F. C. Haber, *The Age of the World: Moses to Darwin*. Johns Hopkins Press, Baltimore, 1959.

A. R. Hall, *The Scientific Revolution 1500–1800*. Beacon Press, Boston, 1962.

A. R. Hall, *From Galileo to Newton*. Dover Publications, New York, 1981.

E. T. Hall, *The Hidden Dimension*. Doubleday, Garden City, NY, 1969.

E. Hamilton, *Mythology: Timeless Tales of Gods and Heroes*. New American Library, New York, 1969.

E. Hansot, *Perfection and Progress: Two Modes of Utopian Thought*. M.I.T. Press, Cambridge, Massachusetts, 1974.

E. R. Harrison, "The natural selection of universes containing intelligent life." *Quarterly Journal Royal Astronomical Society*, **36**, 193, 1995.

E. R. Harrison, *Cosmology: The Science of the Universe*, 2nd edition. Cambridge University Press, New York, 2000.

J. B. Hartle, in *Gravitation and Relativity*, editors N. Ashby *et al.* Cambridge University Press, New York, 1990.

C. H. Haskins, *The Rise of Universities*. Cornell University Press, Ithaca, 1957.

S. W. Hawking, *Is the End in Sight for Theoretical Physics?* Cambridge University Press, Cambridge, 1980.

T. Heath, *Aristarchus of Samos: The Ancient Copernicus*. Clarendon Press, Oxford University Press, Oxford, 1913.

W. Heisenberg, *Physics and Philosophy: The Revolution in Modern Physics*. Harper and Row, New York, 1958.

L. J. Henderson, *The Fitness of the Environment*. Macmillan, New York, 1913.

S. K. Heninger, *The Cosmographical Glass: Renaissance Diagrams of the Universe*. Huntington Library, San Marino, CA, 1977.

K. Heur, *The End of the World: A Scientific Inquiry*. Victor Gollancz, London, 1953.

J. Hick, editor, *The Existence of God*. Macmillan, New York, 1964.

N. P. Hickerson, *Linguistic Anthropology*. Holt, Rinehart and Winston, New York, 1980.

C. H. Hinton, *Scientific Romances*, 2 vols. Swann Sonnenschein, London, 1884–1885.

J. A. Hobson, *Consciousness*. W. H. Freeman, New York, 1999.

B. Hoffman, *Albert Einstein: Creator and Rebel*. Viking Press, New York, 1972.

D. R. Hofstadter, *Gödel, Escher, and Bach: An Eternal Golden Braid*. Vintage Books, New York, 1979.

G. Holmes, *The Later Middle Ages 1272–1485*. Norton, New York, 1962.

S. Hook, editor, *Determinism and Freedom in the Age of Modern Science*. Macmillan, New York, 1961.

R. Hope, *Aristotle's Physics*. University of Nebraska Press, Lincoln, 1961.

M. A. Hoskin, *William Herschel and the Construction of the Heavens*. Science History Publications, New York, 1963.

M. A. Hoskin, "The cosmology of Thomas Wright of Durham." *Journal for the History of Astronomy*, **1**, 44, 1970.

W. Howells, editor, *Ideas on Human Evolution: Selected Essays*. Harvard University Press, Boston, 1962.

W. Howells, *Evolution of the Genus Homo*. Addison-Wesley, Reading, MA, 1973.

D. Hume, *Dialogue Concerning Natural Religion*, editor J. V. Price. Oxford University Press, New York, 1976.

C. Huygens, *The Celestial Worlds Discovered*. Frank Cass, London, 1968. First published in 1690.

D. A. Hyland, *The Origins of Philosophy*. Putnam, New York, 1973.

A. Hyman and J. J. Walsh, editors, *Philosophy of the Middle Ages*. Hackett Publishing Company, Indianapolis, 1973.

W. James, *A Pluralistic Universe*. Macmillan, New York, 1909.

P. James, *Population Malthus: His Life and Times*. Routledge and Kegan Paul, Boston MA, 1979.

M. Jammer, *Concepts of Space*. Harper and Brothers, New York, 1960.

M. Jammer, *The Conceptual Development of Quantum Mechanics*. McGraw-Hill, New York, 1966.

K. Jaspers, *Anselm and Nicholas of Cusa*, editor H. Arendt. Harcourt Brace Jovanovich, New York, 1966.

K. Jaspers, *Anselm and Nicholas of Cusa*. Harcourt Brace Jovanovich, New York, 1974.

J. Jeans, *Scientific Progress*. Macmillan, New York, 1936.

F. F. R. Johnson, "Gresham College: Precursor of the Royal Society." *Journal of the History of Ideas*, **1**, 413, 1940.

P. Johnson, *A History of Christianity*. Atheneum, New York, 1979.

R. Johnson and S. V. Larkey, "Digges, the Copernican System, and the idea of the infinity of the universe in 1576." *Huntington Library Bulletin*, Harvard University Press, April 1934.

F. F. Jones, *Ancients and Moderns: A Study of the Rise of the Scientific Movement in Seventeenth-Century England*. Dover Publications, New York, 1982.

Julian of Norwich, *Revelations of Divine Love*, translated into modern English by C. Wolters. Penguin, Baltimore, 1966.

I. Kant, *A Universal History and Theory of the Heavens*, translator W. Hasties. University of Michigan Press, Ann Arbor, 1969.

E. Kasner and J. Newman, *Mathematics and the Imagination*. Simon and Schuster, New York, 1952.

J. Kidd, *The Bridgewater Treatises, on the Power, Wisdom, and Goodness of God, as Manifested in the Creation, on the Adaptation of External Nature to the Physical Condition of Man*, 6th edition. H. G. Bohn, London, 1852.

G. S. Kirk, *Myth, Its Meaning and Functions in Ancient and Other Cultures*. University of California Press, Berkeley, 1973.

C. S. Kirk and J. E. Raven, *The Presocratic Philosophers*. Cambridge University Press, Cambridge, 1960.

A. Koestler, *The Watershed: A Biography of Johannes Kepler*. Doubleday, Garden City, NY, 1960.

A. Koyré, *From the Closed World to the Infinite Universe*. Harper and Row, New York, 1958.

A. Koyré, *Newtonian Studies*. Chapman and Hall, London, 1965.

H. Kramer and J. Sprenger, *The Malleus Maleficarum*, translator M. Summers. Dover Publications, New York, 1971.

T. S. Kuhn, *Copernican Revolution: Planetary Astronomy in the Development of Western Thought*. Harvard University Press, Cambridge, 1957.

T. S. Kuhn, *The Structure of Scientific Revolutions*, Vol. 2, *International Encyclopedia of Unified Science*. University of Chicago Press, Chicago, 1962.

J. Laird, *Theism and Cosmology*. The Philosophical Library, 1942, reprinted by Books for Libraries Press, Freeport, NY, 1969.

G. Leff, *Medieval Thought*. Penguin, Harmondsworth, Middlesex, UK, 1956.

G. Lemaitre, *The Primeval Atom*. Van Nostrand, New York, 1951.

C. S. Lewis, *The Discarded Image: An Introduction to Medieval and Renaissance Literature*. Cambridge University Press, New York, 1964.

D. C. Lindberg, editor, *Science in the Middle Ages*, University of Chicago Press, Chicago, 1978.

A. O. Lovejoy, *The Great Chain of Being*. Harvard University Press, Cambridge, MA, 1936.

Lucretius, *The Nature of the Universe*, translated by R. Latham. Penguin, Harmondworth, Middlesex, UK, 1951.

M. Maimonides, *The Guide for the Perplexed*, translator M. Friedlander. Dover Publications, New York, 1956.

H. Margenau, *The Nature of Physical Reality: A Philosophy of Modern Physics.* McGraw-Hill, New York, 1950.

M. Marwick, editor, *Witchcraft and Sorcery.* Penguin, Harmondsworth, Middlesex, UK, 1970.

F. Mason and J. A. Thomson, *The Great Design: Order and Progress in Nature.* Duckworth, London, 1934.

E. Mayr, "Darwin and natural selection." *American Scientist,* **65**, 321, 1977.

J. McCann, editor, *Cloud of Unknowing and Other Treatises*, Newman Press, Westminster, Maryland, 1952.

P. McCorduck, *Machines Who Think: A Personal Inquiry into the History and Prospects of Artificial Intelligence.* W. H. Freeman, San Francisco, 1979.

G. R. Montgomery, *Leibniz: Basic Writings.* Open Court, La Salle, Illinois, 1962.

A. Mostowski, *Sentences Undecidable in Formalized Arithmetic: An Exposition of the Theory of Kurt Gödel.* North-Holland, Amsterdam, 1957.

M. K. Munitz, *Space, Time and Creation.* Free Press, Glencoe, IL, 1957.

M. K. Munitz, *Theories of the Universe.* Free Press, New York, 1965.

E. Nagel and J. R. Newman, *Gödel's Proof.* New York University Press, New York, 1958.

J. Napier, *The Roots of Mankind.* Harper and Row, New York, 1973.

J. Needham, editor, *Science, Religion and Reality.* George Braziller, New York, 1955.

O. Neugebauer, *The Exact Sciences in Antiquity.* Brown University Press, Providence, Rhode Island, 1957.

J. D. North, *The Measure of the Universe: A History of Modern Cosmology.* Clarendon Press, Oxford, 1964.

P. H. Nowell-Smith, "Religion and morality," in *Encyclopedia of Philosophy* vol. 7, p. 150. Macmillan, New York, 1967.

L. E. Orgel, *The Origins of Life: Molecules and Natural Selection.* Wiley, New York, 1973.

D. Overbye, *Lonely Hearts of the Cosmos: The Scientific Quest for the Secret of the Cosmos.* Harper, New York, 1991.

S. E. Ozment, editor, *The Reformation in Medieval Perspective.* Quadrangle Books, Chicago, 1971.

S. R. Packard, *12th Century Europe: An Interpretive Essay.* University of Massachusetts Press, Amherst, 1973.

H. R. Pagels, *The Cosmic Code: Quantum Physics as the Language of Nature.* Simon and Schuster, New York, 1982.

H. R. Pagels, *Perfect Symmetry: The Search for the Beginning of Time.* Bantam Books, New York, 1986.

A. Pais, *Subtle is the Lord.* Clarendon Press, Oxford, 1982.

A. Pais, *Inward Bound.* Oxford University Press, Oxford, 1994.

W. Paley, *Natural Theology,* editor F. Ferre. Bobbs-Merrill, New York, 1963.

A. Pannekoek, *A History of Astronomy.* Interscience, New York, 1961.

C. M. Patton and J. A. Wheeler, "Is physics legislated by cosmology?" in *Quantum Gravity: An Oxford Symposium,* editors C. J. Isham, R. Penrose, and D. W. Sciama, p. 150. Clarendon Press, Oxford, 1975.

O. Pedersen and M. Pihl, *Early Physics and Astronomy: Historical Introduction.* American Elsevier, New York, 1974.

W. Penfield, *The Mystery of the Mind.* Princeton University Press, Princeton, 1975.

Plato, *Timaeus,* in *The Collected Dialogues of Plato,* editors E. Hamilton and H. Cairns. Pantheon, New York, 1961.

E. A. Poe, *Eureka,* reproduced in *The Science Fiction of Edgar Allen Poe,* editor H. Beaver. Penguin Books, Harmondsworth, Middlesex, UK, 1976.

C. Ponnamperuma, *The Origins of Life.* Dutton, New York, 1972.

K. R. Popper, *The Logic of Scientific Discovery.* Harper and Row, New York, 1965.

K. R. Popper, "Conjectures and refutations," in *Science: Men, Methods, Goals,* editors B. A. Brody and N. Capaldi. Benjamim, New York, 1968.

V. Postrel, *The Future and its Enemies.* Free Press, New York, 1998.

P. Radin, *Primitive Religion, Its Nature and Origin.* Dover Publications, New York, 1957.

P. Radin, *Primitive Man as Philosopher.* Dover Publications, New York, 1957.

M. Rees, "The collapse of the universe: An eschatological study." *Observatory,* **89,** 193, 1969.

M. Rees, *Our Cosmic Habitat.* Princeton University Press, Princeton, 2001.

R. S. Richardson, *The Star Lovers.* Macmillan, New York, 1967.

W. O. Roberts, "Science, a wellspring of our discontent." *American Scientist,* **55,** 3, 1967.

J. Robson, *Origin and Evolution of the Universe: Evidence for Design.* McGill-Queens University Press, Montreal, 1988.

C. A. Ronan, *Edmond Halley: Genius in Eclipse.* Doubleday, Garden City, NY, 1969.

J. Rosen, "Self-generating universe and many worlds." *Foundations of Physics,* **21,** 977, 1991.

L. Rosenfeld, "Niels Bohr's Contribution to Epistemology." *Physics Today*, October 1963.

R. Rucker, Editor, *Speculations on the Fourth Dimension: Selected Writings of Charles Hinton*. Dover Publications, New York, 1980.

B. Russell, *Human Knowledge, Its Scope and Limits*. Allen and Unwin, London, 1948.

B. Russell, *Why I Am Not a Christian*. Simon and Schuster, New York, 1957.

B. Russell, *The Problems of Philosophy*. Oxford University Press, London, 1959.

B. Russell, *A History of Western Philosophy*. Simon and Schuster, New York, 1972.

M. D. Sahling, *Tribesmen*. Prentice-Hall, Englewood Cliffs, New Jersey, 1968.

S. Sambursky, *The Physical World of the Greeks*. Routledge and Kegan Paul, London, 1963.

N. K. Sanders, *The Epic of Gilgamesh*. Penguin, Harmondsworth, Middlesex, UK, 1972.

E. Sapir, *Language: An Introduction to the Study of Speech*. Harcourt, Brace and World, New York, 1921.

G. Sarton, *Introduction to the History of Science*, Vol. 1, *From Homer to Omar Khayyám*; Vol. 2, *From Rabbi ben Ezra to Roger Bacon*; Vol. 3, *Science and Learning in the Fourteenth Century*. Williams and Wilkins, Baltimore, MD, 1927.

G. Sarton, *Ancient Science: Through the Golden Age of Greece*. Oxford University Press, London, 1953.

J. L. Saunders, *Greek and Roman Philosophy after Aristotle*. Free Press, New York, 1966.

P. A. Schilpp, editor, *Albert Einstein: Philosopher–Scientist*. Library of Living Philosophers, Evanston, IL, 1949.

R. Schlegel, *Inquiry into Science: Its Domain and Limits*. Doubleday, Garden City, NY, 1972.

H. J. Schoeps, *The Religions of Mankind: Their Origin and Development*. Doubleday, New York, 1966.

E. Schrödinger, *What is Life? The Physical Aspect of the Living Cell*. Cambridge University Press, New York, 1946.

E. Schrödinger, *Mind and Matter*. Cambridge University Press, Cambridge, 1958.

E. Schrödinger, *Science, Theory, and Man*. Dover Publications, New York, 1957.

A. Schuster, "Potential matter – a holiday dream." *Nature*, **58**, 367, 1898.

D. W. Sciama, "The ether transmogrified." *New Scientist*, February 2, 1978.

J. B. Sidgwick, *William Herschel, Explorer of the Heavens*. Faber and Faber, London, 1955.

G. G. Simpson, *The Meaning of Evolution*. Yale University Press, New Haven, 1949.

D. W. Singer, *Giordano Bruno, His Life and Thought*. Henry Schuman, New York, 1950.

J. J. C. Smart, editor, *Problems of Space and Time*. Macmillan, New York 1964.

L. Smolin, *The Life of the Cosmos*. Oxford University Press, Oxford, 1997.

D. J. de Solla Price, *Science Since Babylon*. Yale University Press, New Haven, 1961.

B. C. Sproul, *Primal Myths: Creating the World*. Harper and Row, New York, 1979.

O. Stapledon, *Last and First Men and Star Maker*. Dover Publications, New York, 1968.

W. E. Steinkraus, editor, *New Studies in Berkeley's Philosophy*. University Press of America, Washington, DC, 1981.

G. S. Stent, "Limits to the scientific understanding of man." *Science*, **187**, 1052, 1975.

L. Stephen, *History of English Thought in the Eighteenth Century*, 2 vols. Smith and Elder, London, 1881.

B. Stewart and P. G. Tait, *The Unseen Universe*. Macmillan, London, 1886.

J. W. N. Sullivan, *The Limitations of Science*. Viking, New York, 1933.

G. E. Swanson, *Birth of the Gods: The Origin of Primitive Beliefs*. University of Michigan Press, Ann Arbor, 1964.

S. Toulmin and J. Goodfield, *The Fabric of the Heavens: The Development of Astronomy and Dynamics*. Harper and Row, New York, 1961.

H. R. Trevor-Roper, *The European Witch-Craze of the Sixteenth and Seventeenth Centuries and Other Essays*. Harper and Row, New York, 1969.

A. G. Van Melsen, *From Atomos to Atom*. Harper, New York, 1960.

B. L. Van der Waerden, *Science Awakening*, translator A. Dresden. Noordhoff, Gröningen, The Netherlands, 1954.

G. Vlastos, *Plato's Universe*. University of Washington Press, Seattle, 1975.

N. Wade, editor, *Language and Linguistics*. A Science Times Book. Lyons Press, New York, 2000.

A. W. Watts, *Myth and Ritual in Christianity*. Vanguard Press, New York, 1954.

A. W. Watts, *The Two Hands of God*. Collier Books, New York, 1969.

S. Weinberg, *The First Three Minutes: A Modern View of the Origin of the Universe*. Basic Books, New York, 1977.

S. Weinberg, *Dreams of a Final Theory*. Pantheon, New York, 1993.

R. Weisman, *Witchcraft, Magic, and Religion in 17th-century Massachusetts*. University of Massachusetts Press, Amherst, 1984.

V. F. Weisskopf, "Is physics human?" *Physics Today*, June 1976.

R. S. Westfall, *The Construction of Modern Science: Mechanism and Mechanics.* Cambridge University Press, New York, 1977.

R. S. Westfall, *Never at Rest: A Biography of Isaac Newton.* Cambridge University Press, New York, 1981.

H. Weyl, *Space–Time–Matter.* Dover, New York, 1952. Translated from the 1921 German edition.

H. Weyl, *Philosophy of Mathematics and Natural Science.* Princeton University Press, Princeton, 1949.

J. A. Wheeler and K. Ford, *Geons, Black Holes, and Quantum Foam.* Norton, New York, 1998.

L. White, *Medieval Technology and Social Change.* Oxford University Press, New York, 1964.

A. N. Whitehead, *Science and the Modern World.* Macmillan, London, 1925.

A. N. Whitehead, *The Adventure of Ideas.* Free Press, New York, 1933.

W. D. Whitney, *The Life and Growth of Language.* Dover Publications, New York, 1979. First published in 1875.

G. J. Whitrow, *The Natural Philosophy of Time,* second edition. Nelson, London, 1980.

E. Whittaker, *Space and Spirit: Theories of the Universe and the Arguments for the Existence of God.* Thomas Nelson and Sons, London, 1946.

E. T. Whittaker, *The Beginning and End of the World.* Oxford University Press, London, 1942.

E. O. Wilson, *On Human Nature.* Harvard University Press, Cambridge, Massachusetts, 1978.

T. Wright, *An Original Theory or New Hypothesis of the Universe.* Reprint of 1750 edition with an introduction by M. A. Hoskin. American Elsevier, New York, 1971.

Index